Dynamics of Rational Negotiation

Margit Gaffal · Jesús Padilla Gálvez

Dynamics of Rational Negotiation

Game Theory, Language Games and Forms of Life

Springer

Margit Gaffal
Faculty of Commerce and Tourism
Complutense University
Madrid, Spain

Jesús Padilla Gálvez
Facultad de Ciencias Jurídicas
Universidad de Castilla-La Mancha
Toledo, Spain

ISBN 978-3-031-49050-7 ISBN 978-3-031-49051-4 (eBook)
https://doi.org/10.1007/978-3-031-49051-4

This Springer imprint is published by the registered company Springer Nature Switzerland AG
The registered company address is: Gewerbestrasse 11, 6330 Cham, Switzerland

Contents

List of Figures

Chapter 1
Introduction

This volume presents a systematic analysis of the interrelationship that exists between different parties in the negotiation process. In the past, special interest has been placed on studying the formal aspects of the process.[1] In this study, we approach the topic from a new point of view. Our main concern is the interconnectedness between negotiation, game theory, the language games involved in negotiation and the forms of life in which negotiators are embedded. One of the difficulties in understanding the meaning of the technical terms used in negotiations is that they originated in distinct traditions, resulting in different meanings being attached to them. This is a relevant issue that becomes obvious in the following example. In Spanish, French and German, there is no distinction between the words "play" and "game", while in English, there are two clear distinctions. The generic term "play", "jeu" or "Spiel" refers to an original concept in which the notions of "entertainment", "recreational activity" or "unregulated play" are easily confused with the meaning of "game" in English qua physical or mental competition with certain rules that are followed by opposing participants. This difference in meaning is reflected in the history of development that the term has undergone in each tradition.

This semantic and methodological change took place slowly. Inspired by his trips to Paris between 1672 and 1676, Gottfried Leibniz developed an interest in games of luck that were popular in every major capital city at that time. Proof of this is a series of short publications dedicated to the games of "Quinquenove"[2] and "Bassette",[3] respectively. Based on this experience, Leibniz created an academy

[1] Zermelo (1913, 501–504), Borel (1924, 204–224), von Neumann (1928, 295–320), Morgenstern (1935, 337–357), von Neumann and Morgenstern (1944) (2004), Nash (1950a, 48–49), Arrow (1951).

[2] Quinquenove is a historic dice game of luck. A treatise on the game was published in Rémond de Montmort (1708).

[3] The game of Bassette, which took so many victims in France, was invented by the Italian Bassetti and was introduced around 1674 at the French court by Justiniani, ambassador of the Republic of Venice in France.

M. Gaffal and J. Padilla Gálvez, *Dynamics of Rational Negotiation*,
https://doi.org/10.1007/978-3-031-49051-4_1

of games and developed a mathematical theory of board games that can be found in his volume *New Essays on Human Understanding*. In this work, he examined the mathematical foundations of gambling and wrote a critical reply to John Locke's famous philosophical *Essay Concerning Human Understanding*, published in 1690.[4] He pleaded for a new type of logic that would account for degrees of probability and recommended combining it with the analysis of games of luck. He pointed to the need for a mathematical analysis and well-reasoned book on all kinds of games. For Leibniz, the mathematical theory of games of luck is part of the argument he developed against the Lockean approach. In Book IV, Chapter XVI he deals precisely with degrees of agreement in "Des Degrés d'Assentiment" and describes his point of view in the following paragraph:

> "J'ay dit plus d'une fois qu'il faudroit une *nouvelle espèce de Logique*, qui traiteroit des degrés de probabilité [...]. Il seroit bon que celui qui voudrait traiter cette matière poursuive l'examen des *jeux de hasard*; et généralement, je souhaiterais qu'un habile mathématicien voulut faire un ample ouvrage bien circonstancié et bien raisonné sur toute sorte de jeux, ce qui serait de grand usage pour perfectionner l'art d'inventer, l'esprit humain paraissant mieux dans les jeux que dans les matières les plus sérieuses."[5]

To understand what Leibniz's proposition corresponds to, it is necessary to reconstruct the reasoning that led him to formulate it. His approach was based on the question of how we can use our judgement when we have only incomplete knowledge of a situation. The approach proposed by Leibniz differs fundamentally from Locke's reasoning.[6] While Locke considered probability as the appearance of a connection with truth, Leibniz maintained that it stands in a relation of conformity to truth. Where Locke discerned only an accumulation of past experiences practically impossible to remember, Leibniz conjectured an open system in continuous motion in which current beliefs can be revised once new elements appear. Whereas Locke concluded that, because of the difficulties he perceived, it is necessary to decide in full awareness of this uncertainty, Leibniz distinguished between the time of deliberation, in which judgement is exercised on beliefs based on probabilities, and the time of decision, which, once made, must be carried to its conclusion. Leibniz explained his position in the following quotation:

> "Dans nos propres affaires, surtout dans les matières fort importantes, où il est encore permis de s'embarquer et de reculer, et où il n'est pas préjudiciable de suspendre l'exécution et d'aller bride en main, les arrêts de notre esprit, fondés sur des probabilités ne doivent jamais tellement passer *in rem judicatam*, comme les jurisconsultes l'appellent, c'est-à-dire pour établis qu'on ne soit disposé à la révision du raisonnement, lorsque de nouvelles raisons considérables se présentent à l'encontre. Mais quand il n'est plus temps de délibérer, il faut suivre le jugement qu'on a fait avec autant de fermeté que s'il était infaillible, mais non toujours avec autant de rigueur."[7]

[4] Leibniz (1765) (1985), Vol. II. Chapitre XVI, §9, 514.

[5] See footnote 4.

[6] Leibniz (1765), Chapitre XVI. Locke (1690) (1975).

[7] Leibniz (1765) (1985), Vol. II. Chapitre XVI, §3, 500.

In matters of great importance, where a position can be fixed or withdrawn, it is relevant to proceed with tense reins. Leibniz indicated that decisions based on probability should never be valid *in rem judicatam,* a phrase used by jurisconsults. Therefore, we can never maintain the assertion that we are unwilling to review any consideration if significant new reasons are presented against it. Once the process of deliberation has passed, a judgement must be maintained as firmly as if it were infallible, but not always with the same rigour.

What role does the Leibnizian project of a mathematical game theory play in this context? To answer this question, we must discern two fundamental elements: on the one hand, Leibniz saw the need for a systematic treatment of knowledge regarding different degrees of probability, for which games of luck offer an ideal framework. On the other hand, the situation in which the players find themselves being forced to decide is not only representative of this uncertainty but also stimulates their capacity for invention. A theory of board games conceived in this way would therefore allow us to understand how one can make the best use of one's judgement when forced to make a decision in a situation in which one only has access to incomplete information.

Even more precise comparisons can be drawn between Leibniz's outline and the beginning of its elaboration by Borel. In Leibniz's passage in the *New Essays on Human Understanding,* we find the two pillars on which Émile Borel would—more than two centuries later—elaborate a mathematical theory of games based on the "chance and the skill of the players".[8] Chance requires recourse to the calculation of probabilities, while the player's skill intersects with inventiveness and manifests itself in a choice of strategy. Leibniz had argued that in a situation of uncertainty, which characterises most games, once a player has decided on a strategy, he or she must maintain it until an outcome becomes manifest. Leibniz viewed game situations as more conducive to perfecting the art of invention than to the carrying out of any other serious activity. Borel underlined that players not only have to follow the rules of ordinary logic, for instance by eliminating failed strategies but also need to take psychological factors into account. By "psychology", he was referring to the ability to elicit an opponent's strategy and to surprise the other player by replacing the expected rules with other rules invented during the game.[9]

Borel has long been credited with being the first to introduce the concept of a mixed strategy, essential for the solution of two-player zero-sum games. Unlike the so-called pure strategy, a mixed strategy, rather than constituting a specific action, refers to the utilization of a distribution of probabilities across the set of actions available to a player. This definition is not intuitively obvious even today. However, it is by means of this mathematical "trick" that the problem posed by a game in which the players have at least partially different interests could be solved.

A strong objection came from the methodological field and required further clarification.[10] In his work, Ludwig von Mises rejected the assumption that an analogy

---

[8] Borel (1921, 199–203).

[9] Borel (1921, 199–203), Borel (1924, 204–224).

[10] A detailed study evaluating the use and disuse of game theory can be found in Rapaport (1962, 108–119).

can be drawn between games and business. He reduced games to mere skill and cunning and compared the players of a game to tricksters. Games are based on antagonism between two or more opponents, while businesspeople are interested in cooperation.[11] When he studied the behaviour of participants in duopolies and oligopolies, he recognised that negotiators become "rivals". He described participants' actions and reactions as comprising a mutual process of deception where the outcome depends on the opponents' personal shrewdness.[12] He acknowledged that in oligopolies, an opponent's actions may be inconsistent with his or her initial position. This conclusion is in line with similar conjectures made by the linguists J. F. Nash, J. P. Mayberry and M. Subik in their descriptions of duopoly,[13] which suggested that participants' behaviour should be examined from the point of view of game theory.

Evidently, as the scholars mentioned earlier were not familiar with the works of J. von Neumann, O. Morgenstern, John F. Nash, published in the journal *Annals of Mathematics*, they were also unaware of the type of game they were assessing in their evaluation, namely the "non-cooperative game".[14] They also denied that bargaining would be based on so-called game theory,[15] which is expressed by formalised structures in mathematical terms. Without being aware of it, they were applying the so-called zero-sum game,[16] which involves two sides and in which the result is an advantage for one side and an equivalent loss for the other. In other words, the gain of one player is equivalent to the loss of the other, so the net benefit of the game equals zero. This behaviour is based on an antagonistic view of social relations, in which individuals compete for a finite amount of goods in the world, where a person's gain produces a loss for the other person. This primacy of individual interests promotes zero-sum fallacies and generates decisions against the general interest. Consequently, any society that acts in antagonistic games is moving towards disintegration.[17] The present volume aims to explain the dynamics of games, to shed light on unanswered questions and to explain the problems involved.

To date, negotiation proposals have been based on a reductionist-mechanistic model according to which negotiators are viewed as poker players or military strategists with certain psychological characteristics. In this artificial "world", the culture they belong to, the language they speak, the traditions to which they are linked, the ways they behave, etc. seem to play a secondary role when it comes to doing business or interacting with each other. Negotiation is roughly understood as the procedure of generating profits and obtaining maximum benefit. These realities are often ignored, despite the fact that they tend to be a source of misunderstandings that lead to failed

---

[11] Von Mises (2011, 140).

[12] Von Mises (2011, 437ss).

[13] Nash (1996, 47–60).

[14] Von Neumann and Morgenstern (1944) (2004, xii–xiii). Nash (1950a, b, 157–158). Nash (1996, 3–4).

[15] The compound "game theory" was coined by von Neumann and Morgenstern (1944) (2004) and Nash (1996).

[16] Neumann and Morgenstern (1944) (2004, 220ss).

[17] Von Mises (2011, 141).

negotiations. In fact, there is a significant number of problems addressed in the framework of the so-called Nash equilibrium that do not receive appropriate attention. The authors of the present volume focus on negotiators' behaviours, the language games used by negotiators and the forms of life to which they are linked as well as the background and training that influence negotiations.

All the elements mentioned above are relevant for decision-making. If we consider negotiation as a general system of actions, we observe certain recurring patterns of behaviour that can be systematised within game theory. Additionally, we must also consider the domain of language games and the forms of life within which the players operate.[18] It seems that the social system has been left unthematised in game theory; therefore we think that it should be extended by taking these elements into account.[19] We want to explain why relating game theory to language games and forms of life is important for the analysis of real-world negotiations. It may seem somewhat puzzling to read that game theory, which has been extensively and robustly studied, would need to be restructured when attempting to use its results in the analysis of negotiation. However, it turns out that game theory, in the form in which it currently exists, leaves several problems unanswered. In fact, the problem of bargaining as it is addressed in these pages, i.e., the problems that arise, have hardly been studied in their own right and their social, economic, political and even human significance. And, as we shall see, models of cooperative or non-cooperative games offer no explanation as to why certain negotiators choose one game or the other. Nor has due attention been paid to why psychosocial factors influence this decision. Moreover, we do not know, if bargaining is studied from a purely formal point of view and what kind of language games are applied to successfully obtain a profit.

In short, game theory leaves the question unanswered as to why a negotiator tends to choose one strategy and not another. Nor does it provide any insight into the framework in which bargaining takes place and how participants' forms of life affect bargaining itself. In this volume, we focus on expanding this field of issues, which has been left unsketched in classical game theories and therefore remains unresolved in the interaction between negotiators. All these issues, we believe, have an impact on negotiation and its ends and are an important source to be considered when assessing the results of negotiations. Game theorists have focused on the analysis of formal structures, completely neglecting other elements that have generated failures in their proposals. For instance, in the work of John von Neumann and Oskar Morgenstern, the problem of "bluffing"[20] is addressed as an obstructionist phenomenon that impedes negotiation. In his article entitled "The Bargaining Problem",[21] J. Nash also faces a major problem in solving this type of bargaining, since psychological elements come into play, but also a lack of "background" knowledge, which creates

---

[18] Gaffal (2022a, 101–110).

[19] We disagree with A. Rapaport's view of game theory as a "science of conflict" as outlined in Rapaport, 1962, 108. The reason is that according to this view, game theory would inevitably lead to a dilemma and create a "tit for tat" relationship among individuals. Milinski (1987, 433–435).

[20] Neumann and Morgenstern (1944) (2004, 186ss).

[21] Nash (1950a, b, 155–162), Nash (1996, 1–8).

certain disadvantages when it comes to formalising bargaining in negotiations. The intentional overlooking of elementary factors has had some unfortunate effects on the development of bargaining theory: first, game theory has been influenced by the parameters of "win" and "lose" so that formally developed models have been based on simple paradigms—mainly involving games of luck—that do not provide an adequate explanation of the interrelationship between negotiators under more complex conditions. Second, negotiations are viewed as simplified games, drawn via an analogy to gambling relationships in which "game" is considered a formal procedure governed by exact rules that can be described by an axiomatic system. In doing so, the term "gambling" has been homogenised. In these circumstances we need to examine games from a pragmatic perspective[22] and introduce language games so as to extend their content.[23]

As we will see below, the attempts made by J. von Neumann, O. Morgenstern, John F. Nash and other researchers partially fail because the study of bargaining from an exclusively formal point of view does not elucidate many of the processes underlying bargaining. This is because the concept of "game" is extrapolated to the bargaining problem, with the consequence that the procedural aspect is less visible. Therefore, our research sets out to describe the whole process of negotiation. It should be borne in mind that many negotiations are aimed at creating a negotiation framework and that their end is to make contact, to generate an agenda or to find common interests so that negotiations can be conducted in future. In most cases, such preliminary negotiations in which the parties' intentions are to be determined remain unaddressed in game theory because most of the proposed models ignore the assumptions presented in the prenegotiations prior to the game itself.

Since this is an important aspect of negotiation, what must be considered from the outset is that mere participation in negotiations does not give rise to a binding obligation, as in the case of games. For instance, a letter of intent in which an agreement is outlined between two or more parties before an agreement is formalised plays an important role in preliminary negotiations. An invitation to negotiate is nothing more than an expression of a party's willingness to exchange interests. Yet these preliminary processes are not fully represented in mathematical models, as researchers are mainly habituated to approaching the study from a purely formal point of view. In fact, greater importance has been given to two relevant factors to be considered when studying the dynamics of games: the objective of the negotiation and who makes the decisions in the negotiation. In both cases, a distinction has been made as to whether the subject is exclusively one or whether participants are involved. One-person decisions follow very different strategies compared to team decisions, in which multiple objectives play a role. The study of objectives and negotiators'

---

[22] Levinson had developed a special interest in "Pragmatics" taking up the project designed by Wittgenstein when he applied the principle of "meaning is use" (Wittgenstein 2021, §43) to his investigations. He put it like this: "…the well-known slogan "meaning is use" and the insistence that utterances are explicable only in relation to the activities, or language games, in which they play a role". Levinson (1997, 227).

[23] Padilla Gálvez and Gaffal (2013a, 2013b).

strategies cannot be reduced to models based on idealised and univocal formulas that tend to distort the complex process through reductionist simplification.

Traditional game theory has been based on the maximisation of rational deliberation. Individual actions of negotiators are rational if they maximise their payoffs. Negotiators try to get what they value most. The maximisation argument rests on the assumption that the interaction will be rational if the outcome is beneficial to one of the participants. The bargaining objective views the negotiator's action as a means to achieve his objectives in accordance with his preferences. In this volume, we argue that the maximisation theory is not optimal for all cases. Vilfredo F. Pareto showed that a state is optimal if no alternative state can be produced that improves the welfare of at least one participant without reducing the welfare of the others. In short, rational choice interpreted as maximising choice is not optimal for all cases.

This book deals with rational negotiation. By rationality, we mean the ability to negotiate according to consistent principles that satisfy some objective or purpose. Negotiators behave rationally if their decisions are based on choices that result in the optimal level of benefit or utility. Most economic theories assume that all individuals involved in a negotiation behave rationally. The first chapter introduces the concept of negotiation and discusses the most important views. Then the authors explore game theory and contrast cooperative games to non-cooperative games. Finally, the language games involved in any negotiation are analysed. By "language game", we mean any expression that occurs in a specific situation of a negotiation. This includes knowing how two negotiators properly express themselves to reach an agreement, as well as knowing how to reach the best possible agreement in a given situation.

The second chapter focuses on the analysis of bargaining within the framework of game theory. For this purpose, competition, i.e., zero-sum games, is examined. This is followed by a discussion of cooperative games or non-zero-sum games. The classical approach is based on the premise that rational bargaining means making the best decisions for maximising one's own interests. The maximisation approach is criticised and a solution in favour of optimisation will be proposed instead. The so-called Nash equilibrium is described, and mixed strategies are discussed. Finally, the role of bluffing and lying in bargaining is explored. In the third chapter, we introduce a new approach to bargaining as it is studied with different forms of life in mind. First, the concept of forms of life is defined and distinguished from lifestyle. Next, the core elements of the concept are examined. The authors go beyond the standard two-sided negotiation in which both negotiators have the same cultural background but also examine the variety of scenarios and contexts in which to negotiate rationally with multiple opponents, problems and constraints. Finally, some objections are addressed and unresolved problems in the research on negotiation are presented.

The fourth chapter focuses on the psychological aspects of negotiation. The authors analyse common mistakes linked to negotiator behaviour. First, the conditions under which negotiation takes place are examined, after which persuasive communication is analysed. Subsequently, the legal consequences of deception are outlined. Tricks in negotiation are systematised and the role of body language and deception is studied. The fifth chapter introduces mediation in the field of negotiation. Thus, mediation is defined within the framework of game theory. The suitability of negotiators for mediation is then explored. Subsequently, the characteristic elements of conflict in negotiation are presented. This chapter ends by exploring the concepts and models developed in mediation.

This book aims to illustrate some of the most important elements to be considered when studying the negotiation process. Our research has benefited from different sources. On the one hand, the research carried out jointly on language games and forms of life. The origin of these terms lies in the studies carried out by the linguist Karl Bühler in his "*Theory of Language*"[24] and are now being applied by linguists to game theory. It has also drawn on comparative studies between different cultures when assessing interference in communication. The results in this book are based on Prof. M. Gaffal's experience teaching two courses on negotiation in English at Complutense University in Madrid over nine years. We would like to take this opportunity to thank her students for their suggestions and the challenges they raised during these years. Some of these suggestions have influenced the approach to this work. Although this work is published jointly, some of the chapters have been analysed and dealt with more intensively by one of the authors. Overall, we hope that it will encourage further research into a highly complex and fascinating field in which the last word has yet to be said.

References

Arrow, Kenneth J. 1951. *Social choice and individual values*. New York: Wiley.

Borel, Émile. 1921. La théorie des jeux et les équations à noyau symétrique gauche. *Comptes rendus de l'Académie des sciences, CLXXIII*, 1304–1308, In: Borel, 1924b, 199–203.

Borel, Émile. 1924. Sur les jeux où interviennent le hasard et l'habilité des joueurs, en: *Eléments de la théorie des Probabilités* 204–224. Paris: Librairie Scientifique, ed. by J. Hermann.

Bühler, Karl. 2011. *The theory of language. The representational function of language* 62–63, 220. Amsterdam: John Benjamins Publishing Company.

Gaffal, Margit. 2022. Negociación, teoría de los juegos, juegos de lenguaje y formas de vida. *Dókos* 29–30: 101–110.

Leibniz, Wilhelm Gottfried, 1765. *Nouveaux essais sur l'entendement humain, Neue Abhandlung über den menschlichen Verstand*. Darmstadt: WB, 1985, 500, 514.

Levinson, Stephen C. 1997. *Pragmatics*, 227. Cambridge: Cambridge University Press.

Locke, John. 1690. *An essay concerning human understanding*. Oxford: Clarendon Press, 1975.

Milinski, Manfred. 1987. TIT FOR TAT in sticklebacks in the evolution of cooperation. *Nature* 325: 433–435.

Morgenstern, Oscar. 1935. Vollkommene Voraussicht und wirtschaftliches Gleichgewicht. *Zeitschrift Für Nationalökonomie* 6 (3): 337–357.

Nash, John F. 1950a. Equilibrium points in n-person games. *Proceedings of the National Academy of Sciences of the United States of America* 36: 1.

Nash, John F. 1950b. The bargaining problem. *Econometrica* 18: 157–158.

Nash, John F. 1996. Essays on game theory. *Cheltenham, Brookfield: Edward Elgar* 3–4: 47–60.

Padilla Gálvez, Jesús, Margit Gaffal, (Eds.) 2013a. *Forms of life and language games* (2nd ed.) Berlin: De Gruyter.

[24] Bühler (2011, 62–63, 220).

Padilla Gálvez, Jesús, Margit Gaffal, (Eds.) 2013b. *Formas de vida y juegos de lenguaje*. Madrid, México D.F.: Plaza y Valdés.

Rapaport, Anatol. 1962. The use and misuse of game theory. *Scientific American* 207 (6): 108–119.

Rémond de Montmort, Pierre, 1708. *Essay d'analyse sur les jeux de hazard*. Paris: Jacque Quillau, Imprimeur-Juré-Libraire de l'Université.

Von Neumann, John, 1928. Zur Theorie der Gesellschaftsspiele, *Mathematische Annalen, 100*, 1, 295–320.

Von Mises, Ludwig. 2011. *La acción humana. Tratado de economía* 140, 141, 437, . Madrid: Unión Editorial.

Von Neumann, John, and Oskar Morgenstern. 1944. *Theory of games and economic behaviour* 186ff, xii–xiii, 220ff. Princeton: Princeton University Press, 2004

Zermelo, Ernst. 1913. Über eine Anwendung der Mengenlehre auf die Theorie des Schachspiels, In *Proceedings of the Fifth International Congress of Mathematicians* vol. II, 501–504. ed. by Ernest W Hobson, and Augustus E. H. Love (Eds.), Cambridge: Cambridge University Press.

Chapter 2
Negotiation, Game Theory and Language Games

1 Introduction

In this first chapter, we aim to challenge a generally accepted assumption that when two people enter negotiation, they coincide in their modes of expression and converge in their forms of life. It is problematic to assume that both negotiators will communicate in the same language, belong to the same social group and share the same beliefs and points of view. Even more, it turns out to be a misleading assumption that dialogue takes place at the same level for both participants and that the interlocutors coincide in form and content. Anyone who has conducted or taken part in a negotiation will easily understand that this ideal model clashes with everyday experience and therefore must be incorrect. If a German businessperson negotiates with an Italian, Argentinean, Canadian, Lebanese or Japanese colleague, the two parties' commands of English as means of communication are different, their knowledge and negotiation strategies have been acquired in different schools, their behaviour depends on their own cultures and forms of life and their particular ways of expressing themselves in the lingua franca are complicated by their mother tongues and so on.

This complexity results from the interaction of many factors and has been described by two scholars: the mathematician John von Neumann and the linguist Karl Bühler. John von Neumann developed the structure of a social game theory in 1928[1] and, together with Oscar Morgenstern, published a book on rule-governed behaviour in social interactions.[2] Karl Bühler published his book *Sprachtheorie. Die Darstellungsfunktion der Sprache*[3] in 1934, in which he presented a theory of combined language games inspired by Wittgenstein's writings.[4] These proposals

[1] Von Neumann (1928, 295–320).

[2] These pages show the embryonic development of game theory and the influence of the Vienna Circle and Karl Menger's Mathematical Circle, cf: von Neumann and Morgenstern (1944) (2004, 714–715).

[3] Bühler (1965, 48ff).

[4] Wittgenstein (2021).

M. Gaffal and J. Padilla Gálvez, *Dynamics of Rational Negotiation*,
https://doi.org/10.1007/978-3-031-49051-4_2

Fig. 1 Chessboard

arose in environments, in which interdisciplinary work generated great advances that had major repercussions, particularly in the field of negotiation research.

Several related issues were incorporated into this study. The first problem to be clarified is what is meant by "game". The origin of the current meaning of "game" traces back to the foundational work of Ernst Zermelo.[5] To understand the term's new meaning, it is helpful to examine a game of chess like the one in Fig. 1.

Zermelo suggests the following considerations: first, we must determine the value of any possible position for each player, seeking the best possible move. As can be seen in the position of the figures on the schematic chessboard, one of the players holds a special position such that he could force checkmate. The relevant question in a game theory is whether this position also applies more generally. This is how the fundamental question of the definition and existence of strictly dominant strategies has been introduced. Evidently, the strategies pursued by both players lead to a finite game, which means that the game is carried out through a finite number of possible positions. Zermelo's most relevant conclusion was the following: in finite games involving two players, such as in chess, either there exists a dominant strategy for one player, such that he can win the game independently of the strategy of the other player, or there is no such strategy.

A few years later, Émile Borel took up some relevant proposals developed by Leibniz, who had argued that games can be characterised as a process of uncertainty. Leibniz concluded that, once the players have determined their positions, they must maintain them until the final outcome. This perspective recognises the technical importance of a strategy, which in game theory corresponds to a complete plan of action that is maintained consistently until the end of the game. Two centuries later, Borel started from Leibniz's somewhat enigmatic estimation that game situations are more conducive to perfecting the art of invention than to negotiating serious activities

[5] Zermelo (1913, 501–504).

and offered a new assessment.[6] What Borel observed in a simplified game was that the players not only have to follow the prescriptions of ordinary logic by eliminating dominant strategies, but such games also require a certain psychological mastery. By "psychology", he meant the ability to observe the way an opponent operates in a game. More specifically, a player must test his opponent's ability to surprise him by replacing consistent rules of behaviour with other unexpected moves invented during the game. Based on the considerations outlined above, we identify two basic elements of game theory that interact in a game: chance and player skills. Chance requires that the player relies on the calculation of possibilities, while a player's skill has to do with ingeniousness and manifests itself in the choice of a strategy.

This procedure was translated into a general game theory in order to describe the behaviour of two or more players. Without describing all possible ways of playing the game of chess, it seems sufficient to focus on those definitions that are most appropriate for a negotiation. John von Neumann and Oscar Morgenstern took up the above intuitions and proffered the following definition:

> First, one must distinguish between the abstract concept of a *game*, and the individual *plays* of that game. The *game* is simply the totality of the rules which describe it. Every particular instance, at which the game is played – in a particular way – from beginning to end, is a *play*.
>
> Second, the corresponding distinction should be made for the moves, which are the component elements of the game. A move is the occasion of a choice between various alternatives, to be made either by one of the players, or by some device subject to chance, under conditions precisely prescribed by the rules of the game. The *move* is nothing but this abstract "occasion," with the attendant details of description – i.e. a component of the *game*. The specific alternative chosen in a concrete instance – i.e. in a concrete *play* – is the *choice*. Thus the moves are related to the choices in the same way as the game is to the play. The game consists of a sequence of moves, and the play of a sequence of choices.
>
> Finally, the *rules* of the game should not be confused with the *strategies* of the players (…) the distinctions which we stress must be clear from the start. Each player selects his strategy – i.e. the general principles governing his choices – freely. While any particular strategy may be good or bad – provided that these concepts can be interpreted in an exact sense (cf. 14.5 and 17.8-17.10)– it is within the player's discretion to use or to reject it. The rules of the game, however, are absolute commands. If they are infringed, then the whole transaction by definition ceases to be the game described by those rules.[7]

As we can see, it is necessary to distinguish the abstract concept of "game"—as "game" is understood in English—from the "individual plays" of the same. Later on, we will stress again the importance of the terms "*game*" and "*play*", which are both equivalents to the English translation of "juego" but whose meaning differs essentially. A game can be characterised by the totality of the rules that describe it.[8] A play is characterised as any particular instance in a game. Furthermore, moves are to be distinguished, since they are considered the elements of a game. A player—or in our case a negotiator—chooses by means of a move between several alternatives

[6] Borel (1924, 204ff).

[7] Von Neumann and Morgenstern (1944) (2004, 49).

[8] Wittgenstein would indicate that "to follow a rule" is, in short, to act in a certain way. Wittgenstein (2021, §143). Cf.: Gaffal (2022, 213ff).

set out in the rules of the game. The move is a component of the game that can be described at any time. The player always chooses a particular alternative in determining a move. Therefore, moves are directly linked to a choice made by any player and the game must be consistent with the move. In short, every game consists of a sequence of moves, and this sequence is determined by a sequence of choices.

Moreover, the rules of a game should not be confused with the strategies developed by the players. Each party freely chooses its strategy, i.e., the general principles governing its choices. It is therefore at the discretion of each player to use or reject a particular strategy. On the contrary, the rules of the game are considered imperative, so that if the rules are violated then the business is, by definition, no longer the game described by those rules.

Evidently, the concept of "game" as approached by von Neumann and Morgenstern is conceived from an abstract perspective. The starting point of the definition presented above is reduced to a description based on a formal model. This abstract model of "game" operates with the minimum of information necessary to arrive at a solution. The actual alternative courses of action from which the negotiators must choose are not taken into account, as they are not considered essential information. These alternatives are treated as abstract objects with no special qualities and are called "strategies". In fact, his proposal only considers players' attitudes towards the final outcomes of the imagined possible strategies.

Game theory is considered a theory of strategic interaction.[9] In other words, it is a theory of rational behaviour in social situations in which each player must choose his moves based on what he thinks the other players are likely to do. Its ultimate goal is to help us define players' rational behaviours in real-life economic, political and social situations. Therefore, we will call the process of optimal decision-making by two or more negotiators with different objectives a *game theory in the framework of bargaining*. Game theory is a mathematical theory that attempts to resolve the conflicts involved in any negotiation. A conflict can be described as an interactive situation in which several negotiators make decisions and in which the outcome depends on each negotiator's preferences over the set of possible outcomes.[10] Due to the reciprocal effect of the preferences of each player, we believe that some important elements can help to explain a real-life negotiation. The model involves rational, psychological and sociological elements that also appear in negotiations. In fact, it is a kind of umbrella theory for the rational side of social science, where "social" is interpreted in a broad sense to include human actors as well as the environment. Its methodologies apply in principle to all interactive situations, especially in economics, political science, evolutionary biology and computer science. There are also important connections with accounting, statistics, the foundations of mathematics, social psychology, law, business and branches of philosophy such as epistemology and ethics.

We start from the fact that all negotiation dialogue is governed by the strategies and tactics developed in game theory. The languages used by negotiators are part of

---

[9] Von Neumann and Morgenstern (1944) (2004, 79f).

[10] Myerson (1991, 1ff).

what are called "language games",[11] which are closely linked to the forms of life[12] of the participants in a negotiation. Negotiations seem to follow a dialogical logic that involves concepts from both game theory and argumentation. The rules of the game allow the players to defend their own interests in such a way that one negotiator can win and the other lose, or both can win, or both can lose. The model outlined will show that one negotiator can win or lose, regardless of the moves of the other negotiator, according to a winning or losing strategy.

The dialogues that take place in the framework of a negotiation are considered finite games revolving around an initial requirement, for instance, the desire to acquire a product under certain conditions (price, quality, etc.) and for specific purposes. The whole process is valid by both parties as long as the negotiator proposing the deal develops a winning strategy.[13] As we will show later, the game may be fictitious in the sense that one of the parties is only interested in obtaining information or does not really intend to establish a business relationship and to conclude a contract. When negotiating an issue, the parties follow certain rules: one party makes an offer, and the other party accepts, asks for modifications, or makes a counteroffer. The term *ophelimity* was first introduced by Vilfredo Pareto to propose a measure that describes the function of economic goods and services in satisfying human wants and needs. When explaining how consumer desires are satisfied, it is relevant to consider the concepts of pleasure, practical benefit, and utility. These three concepts refer to certain subjective characteristics: pleasure is the experience of finding a commodity, action or feeling enjoyable and personally gratifying. A product's utility refers to its ability to meet a specific need. It is important to maintain objectivity and avoid subjectivity when discussing utility. A product's practical benefit refers to its ability to meet a specific need. It is important to maintain objectivity and avoid subjectivity when discussing practical benefit. And finally, utility is considered a general measure of the satisfaction one derives from obtaining a product or receiving a service. To determine the utility of a product or service thus plays a crucial role in the development of a negotiation process. In fact, the negotiating parties must find out what desires or needs a product or service satisfies for the other party to obtain an advantage. The advantage gained by each of the negotiating parties is negotiated by having them compensate each other.[14] Although there may be common interests, both negotiators view themselves as rivals during the negotiation, as they also have partially conflicting objectives. In fact, the dialogues that take place in a negotiation

[11] In his book entitled *Pragmatics*, St. C. Levinson took up the project designed by Wittgenstein, who coined the expression "meaning is use" (Wittgenstein 2021, §43) and insisted that utterances can only be understood in the contexts in which they occur. He put it like this: "…the well-known slogan "meaning is use" and the insistence that utterance are only explicable in relation to the activities, or language-games, in which they play a role". Levinson (1997, 227).

[12] Forms of life represent the unjustifiable basis on which individual language games receive their meaning. For example, "asking questions", "harbouring doubts", "giving orders" or "giving gifts" can only take place within the framework of social customs that are unquestionably recognised both explicitly and implicitly in each society.

[13] Kamlah and Lorenzen (1973, 211–213), Lorenzen and Lorenz (1978, 8, 12f).

[14] Gaffal, Margit. 2023: Ophelimity and the Paradox of the Stag or the Hare. *Dókos*, 31–32, 97–107.

can be understood as a kind of "game of giving and asking for reasons",[15] and this game can generate joint objectives (e.g., the offeror asks the seller to distribute a product in a certain area). There is no unanimity among scholars as to whether or not the framework of allows for cooperation. It can be argued that the original dialogical framework in any negotiation is only oppositional and adversarial and does not involve cooperative games. However, it can also be argued that any dialogue in a negotiation need not be entirely adversarial, since the start of a negotiation may only be a preliminary agreement whereby the parties generate a non-binding document setting out the main issues to be discussed. The expression "invitation to negotiate" is part of contract law and comes from the Latin phrase *invitatio ad offerendum*, meaning "invite an offer". Thus, one of the negotiators makes an invitation to treat and thus expresses a "willingness to negotiate".[16]

The actions performed in a negotiating dialogue are called "moves"[17] and are understood as speech acts involving declarative utterances (statements) and interrogative utterances (requests).[18] A winning strategy can be described in the following dialogue: starting from a reduction method and applying a normal form, we can denote the disjunctions in the following way: $(A \vee B) \wedge C \rightarrow (A \wedge C) \vee (B \wedge C)$. Such a procedure can be dialogically described by the following strategy between an opponent (O) and a proponent (P):

Opponent		Proponent	
		$(A \vee B) \wedge C \rightarrow (A \wedge C) \vee (B \wedge C)$	
$(A \vee B) \wedge C$		?	
$(A \vee B), C$		?	
A	B	$(A \wedge C) \vee (B \wedge C)$	
?	?	$(A \wedge C)$	$(B \wedge C)$
?	?	A, C	B, C

The winning strategy applied by the proponent corresponds to the semantic tableau introduced by Evert W. Beth.[19] As the proponent's strategy is aimed at confirming only those elementary propositions that the opponent has already asserted beforehand, the proponent can win every proposition in this dialogue. However, the proponent may be forced by the opponent to assert an elementary proposition that the opponent has not yet confirmed. Then, the proponent may not be able to win all

[15] Lorenzen and Lorenz (1978, 28).

[16] A negotiator who makes an invitation to treat does not intend to be bound as soon as it is accepted by the person to whom the statement is addressed. If a product is advertised in large quantities in a newspaper or on a poster, it is generally not considered an offer, but will be considered an invitation to treat, since there is no guarantee that the shop can supply the item to all those who might want it. The newspaper advertisement could only be an invitation to treat, as it could not have been intended as an offer to everyone. Similarly, a product display in a shop window is an invitation to treat.

[17] Lorenzen and Lorenz (1978, 4).

[18] Searle (1969).

[19] Beth (1962, 40ff), Lorenzen and Lorenz (1978, 11f).

propositions of the asserted model. He may not be able to prove that the elementary proposition is affirmed. Consequently, in the dialogue system, the "logical validity" of a proposition is defined by the existence of a strategy for the proponent in which he only has to confirm propositions that have already been asserted by the opponent. In such cases, the tableau is said to be closed.

According to the rules of dialogue, utterances are never isolated items but are interpreted as (parts of) speech acts. As such, these rules define the meaning of utterances. This approach to meaning is rooted in Wittgenstein's observation that in language there is no outside perspective to determine what something means and how it is linked to syntax.[20] In other words, language is inevitable, one cannot go beyond it. Consequently, language is studied with and through language games and these language games are part of a negotiation.[21] In this sense, all speech acts related to meaning and expression are made explicit within the dialogical framework. Likewise, the rules that apply in negotiation are considered "*rules of the game*" and constitute the dialogical approach. A distinction must be made here between the strategic level of the negotiators and the goals pursued by each of the parties. The level of the negotiating game consists of the individual moves made by each of the negotiators according to the rules of the game that each one applies individually, while the level of strategy consists of a certain perspective on all the possible moves towards the same end, outlined from the very beginning of the negotiation.

In this first chapter, we connect the concept of negotiation and game theory to language games and forms of life. The nexus by which all these elements are combined is the concept of rule.[22] The concept of "rule" is polysemous, but in this case, it refers to a prescribed procedure that is observed by negotiators and that regulates and controls their actions. For this reason, following a rule has a positive or negative impact on the negotiation depending on the participants' language games and forms of life and whether these generate agreements or conflicts.

2 Negotiation

We have introduced a definition of negotiation that describes a process of discussion between two or more parties, the aim of which is to reach a mutually acceptable agreement.[23] Three conditions are essential for a disagreement to be resolved through

[20] Wittgenstein (2021, § 496f).

[21] Thus, Levinson indicates that an interpretative corollary of language games is encompassed by the notion of inferential schema. That is to say, inferential schema refers to the formal structure that makes it possible to obtain a well-formed expression detached, free, as a theorem from a formal system previously defined by the strict separation rule of formation and transformation. A framework is a knowledge set that is evoked in order to provide an inferential basis for the understanding of an utterance. Cf.: Levinson (1997, 280f).

[22] Gaffal (2022, 213f).

[23] Among the different negotiation models currently in existence, one can compare the present approach with Powell (2012).

negotiation: first, the existence of a minimum of common or complementary interests between the parties; second, the motivation on both sides to reach an agreement; and third, the existence of mutually recognised autonomy. The parties' objectives can be varied, for instance, to resolve points of difference, achieve advantages for a stakeholder, favour outcomes that satisfy various interests, improve current situations or to resolve conflicts.

The rules governing negotiations determine which sequences of moves are made by which negotiators and whether these moves are allowed. The elementary rules determine the permitted moves of a negotiation and regulate the interaction according to how things proceed in proposals, agreements and conflicts.[24] They are appropriate ways of giving and/or receiving justifications for courses of action and are specific to each type of business according to their respective fields. If a buyer asks about the terms and conditions or the characteristics of a product, it is normal for the seller to provide the information available, as long as this is not used to leverage a better offer from a competitor. These initial rules reveal the negotiators' commitments and rights with respect to each piece of information provided and thereby establish in advance which moves are allowed and appropriate in the negotiation game. As such, the rules of negotiation are an integral part of the dialogue and accord with the established business culture in which given negotiators interact.

Apart from the elementary "exit" rules in negotiation, there are also certain structural rules that determine the general course of a negotiation, e.g., how it is initiated, how it proceeds, how to deal with an impasse, how it ends and so on. The purpose of these rules is not to specify how negotiators must act but rather to determine the procedure according to which negotiators interact. It is one thing to determine the subject matter of a negotiation, be it purchase, sale, acquisition, distribution, etc., and quite another to define whose turn it is to negotiate and when a negotiator is allowed to make a move. Indeed, one could change a rule during the negotiation and force one of the parties to make two "moves at once", for instance, to fix the quantity of a product and to include warranty conditions not previously mentioned. All this would change the negotiation considerably, but without negating the overall deal per se.

We now move to a higher level and focus on the moves developed in a negotiation as a whole. More specifically, we examine the strategic level at which we can analyse collective moves made by negotiators rather than remaining at the level of individual moves.[25] Among the main reasons for doing so are the following: we can determine whether a negotiator is solvent, or whether the contract can be properly finalised and executed. As such, it is relevant to know whether a negotiator really intends to negotiate or whether he is just pretending to do so for unstated purposes. The strategies involved in negotiation are studied under the rubric of game theory.

---

[24] A representation can be found in: Schulte and Delgrande (2004, 73–101).

[25] Von Neumann and Morgenstern (1944) (2004, 79–84).

3 Game Theory

Game theory is used to study the interactions that can take place during a negotiation by means of models. To do this, the negotiation is formalised by compiling the possible incentives that may direct the active parties. These incentives are called "games" in the sense that they follow rules. Thus, game theory provides a model for understanding human behaviour in decision-making.[26] It allows us to analyse the strategies that negotiators develop throughout a negotiation as well as to elicit their expected and observed behaviours. When setting up the model, it can happen that different interactions may share similar features and can therefore be represented by the same game. Game theory is applied to bargaining because both positive and negative behaviours can be observed and described in decision-making. Two people can negotiate and may profit from the negotiation while distrusting each other at the same time. These relations and the variety of possible interactions constitute the field of study of game theory. Although based on a mathematical analysis, it involves analysis of the observed interaction or conflict from a rational point of view. According to this approach, a "game" is a conflict situation in which the negotiators' opposing interests prevail. In such a context, any decision made by one negotiator will influence the decision made by the other. The outcome of the negotiation is measured by considering all decisions made by all negotiators.

If we approach game theory from a decisional point of view, then we can analyse one or several objectives in a negotiation.[27] If we relate these to a single decision-maker or to several participants, the field of research becomes extremely complex: on the one hand, we study the behaviour of the single decision and differentiate it from the one in which the decision contains multiple objectives. On the other hand, if we study several negotiators who make decisions within the negotiation framework, we need to distinguish between cooperative and non-cooperative games. The following table shows the possibilities that arise when one or more negotiators each pursue one or more objectives (Fig. 2).

The rational presupposition that underpins game theory can be described as follows: in any negotiation, there must be a rational way to negotiate a conflict, especially when deceptive actions are involved.[28] It enables negotiators to anticipate their opponents' intentions by means of the rational combination of possibilities, facilitating solutions to complex problems. Neumann, Morgenstern and Nash schematized the basic postulates of game theory and modified the way negotiators have to make decisions. They classified games into different categories and described which methods can be applied to solve what ends. These categories include symmetric and asymmetric games,[29] zero-sum and non-zero-sum games,[30] maximin

[26] See the results published in Padilla Gálvez (2016, 59ff).

[27] The so-called "*Entscheidungsproblem*" was solved by Church (1936, 40–41) and Turing, (1936–1937, 230–265).

[28] Padilla Gálvez (2021, 226–232).

[29] Von Neumann and Morgenstern (1944) (2004, 165f).

[30] Von Neumann and Morgenstern (1944) (2004, 85ff, 291ff).

Fig. 2 Possible combinations among one or more negotiators with one or more objectives

	one objective in the negotiation	*several objectives in the negotiation*
One negotiator decides	Decision theory single-person	Decision theory with multiple objectives
Several negotiators decide	Cooperative game theory	Non-cooperative game theory

Decision theory

and minimax criteria,[31] the Nash equilibrium,[32] cooperative games, non-cooperative games,[33] simultaneous and sequential games, perfect information games[34] and many others that are currently being developed. Our interest is to expose some important results that are closely associated with bargaining.

Cooperative models assume that agents use cooperative mechanisms to make binding commitments outside the specified rules of a game. Cooperative solutions focus on how to allocate the benefits resulting from cooperation. In contrast, non-cooperative models assume that players lack cooperative mechanisms or decide not to apply them. Some non-cooperative models analyse strategic games (including finite games, zero-sum games, matrix games, etc.), extensive games (perfect information games, perfect and imperfect memory games, incomplete information games, etc.) and dynamic games (iterative games, differential games, stochastic games, etc.). Among cooperative models, we will analyse problems related to bargaining, in which the participants negotiate an issue under dispute.[35]

4 Cooperative Games

A game is cooperative if the players compete to achieve a certain goal. Cooperation is considered a strategy for finding out which negotiators share a common interest. A cooperative game is played in so-called coalitions, in which certain groups of players create alliances in line with given external rules, for instance, given by contract law. Cooperative game theory also deals with situations in which players can prenegotiate about what to do in the game. It is common to assume that these negotiations culminate in the signing of a binding agreement. Under these conditions, it is argued, the precise strategies available in the game will not matter much. What is more relevant

[31] Von Neumann and Morgenstern (1944) (2004, 154).

[32] Nash (1996, 13–21).

[33] Nash (1996, 48, 51–52).

[34] Von Neumann and Morgenstern (1944) (2004, 112ff).

[35] Nash (1950, 155–162), Nash (1996, 1–8).

is the preferred structure of the game, since it is this structure that determines which contracts are feasible.

In the second part of von Neumann and Morgenstern's book, they focused mainly on studying the formation of coalitions involving many players in cooperative games.[36] They described the stable classes that generate potential outcomes and abandoned the goal of finding a unique solution. They were the first theorists to construct a cooperative theory of n-person games. They started from the assumption that several groups of players can form coalitions, each of which has an associated value defined as the minimum amount that the coalition can secure by its own efforts. In practice, these groups appear to be organised around blocks linked to a legislative body, or business partners in a corporate conglomerate.[37] Neumann and Morgenstern described these n-person games in the form of a characteristic function by enumerating the individual players (as one-person coalitions) of all possible coalitions of two or more players and the values that each of these coalitions could secure if a countercoalition formed by all other players acted to minimise the amount that the coalition could obtain at the end of the game or as a result of the bargaining. They also assumed that the characteristic function was superadditive, which means that the value of a coalition of two previously separate coalitions is at least as large as the sum of the separate values of the two coalitions.[38] They not only studied the optimal strategies[39] but also predicted and observed the behaviour of individuals in the game.

A paradigmatic case was conducted by analysing a problem in which two individuals negotiate how to divide a sum of money among them. If they do not cooperate mutually, they do not receive any money as they cannot reach an agreement. They indicated that two rational negotiators will resolve their differences efficiently and allocate to each negotiator at least as much as they could get by refusing to agree at all. This is because if both cooperate, then they can share some sum of money but if they refuse to cooperate neither of the two receives anything. If one party cooperates and the other refuses, the result will be that one party receives the entire profit and the other receives nothing. However, this would be against the rules and would violate the principle of legal certainty, because one party would unjustifiably enrich itself at the expense of the other.[40] This corollary stands contrary to the legal foundations on which the principle of cooperation is based. The outcomes were referred to as the "bargaining package", and it was argued that to further specify the outcome, information on the "bargaining skill" of both negotiators would be needed (Fig. 3).

John Nash reframed the whole bargaining process in terms of the "bargaining skill" of each party[41] but corrected this expression by indicating that one cannot speak of one rational bargainer being more skilled at bargaining than another, since each will

[36] Von Neumann and Morgenstern (1944) (2004, 221f).

[37] Von Neumann and Morgenstern (1944) (2004, 238ff).

[38] Von Neumann and Morgenstern (1944) (2004, 250f).

[39] Arrow (1983) (1968, 107–132).

[40] Unjustified enrichment is a criminal offence because it undermines legal foundations. In game theory it could be considered a meta-rule and therefore is prohibited.

[41] Nash (1950, 155–162) and Nash (1996, 1–8).

Fig. 3 Payoff matrix in cooperative games

	Cooperate	*Inhibit*
Cooperate	Both gain through mutual cooperation	He who inhibits wins and he who cooperates loses
Inhibit	He who cooperates loses and he who refrains wins	Both lose through mutual inhibition

Cooperative game

simply adopt whatever bargaining technique is optimal in a given situation.[42] As a result, the question arose as to whether bargaining is indeterminate. Nash changed the accepted view by proposing a set of axioms that were meant to provide a rational solution to the bargaining problem. He then showed that the set of axioms admits a unique solution in bargaining. His novelty lay in positing that the fact that the solution of a bargaining problem should be defined as a function covering the set of possible bargaining problems and the entire set of possible outcomes. According to Nash, a bargaining solution is to be understood as the outcome of bargaining between two rational negotiators who are concerned exclusively with receiving as large a fraction of the bargained-over content as possible.[43] However, one of the axioms proposed by Nash explicitly excluded comparison of the utility accrued by each player when applying the Nash bargaining solution. The problem caused by this proposal was this: how can we clarify to what extent it is possible to speak of fairness when it is impossible to compare who gets how much in a negotiation?

As already indicated, cooperative game theory presupposes a pregame bargaining period during which the players reach an agreement on how to bargain. However, all prenegotiation is itself a particular kind of game. The moves are the proposals and statements that the negotiators exchange in the prenegotiation, and the outcomes are the ways in which the negotiation will be conducted, which suggests at the outset that the parties are willing to reach an agreement. If all the moves made during prenegotiation and negotiation are specified, then we would have a complete game theory. This "complete" game has to be studied without presupposing any pregame negotiations, since the prenegotiations have already been incorporated into the rules. From this point of view, studying a game without presupposing any prenegotiation amounts to proposing a non-cooperative analysis, which would naturally start by examining all Nash equilibria of the game. Consequently, if we were to solve the equilibrium selection problem by means of a non-cooperative analysis of this game in non-cooperative bargaining, we could solve any problem in cooperative game theory. Therefore, to describe the cooperative solution of a game, we would have to include in the negotiation also the rules prefixed in the prenegotiation.

[42] Nash (1953, 128–140) and Nash (1996, 34–46).

[43] Nash (1953, 130) and Nash (1996, 36).

Clearly, describing all the strategies that negotiators resort to before, during and after a negotiation is so complex that its formalisation would require multiple factors to capture the elements involved in that process. But at least the attempt to do so would be extremely ambitious. In fact, the models of cooperative and non-cooperative games described so far have to be considered as complementary since a negotiator can cooperate up to a certain point and then change strategies for unknown and even untoward reasons. However, we know from Nash's research that the use of cooperative solution concepts is more viable when it comes to predicting the outcome of a negotiation. The reason for using a cooperative solution concept for prediction purposes, rather than analysing a non-cooperative model of actual negotiation procedure, is that the latter must necessarily incorporate all sorts of details about which the negotiator is unlikely to be informed, and that much of this information will be irrelevant to the final outcome.[44] Demanding a cooperative solution concept to predict the outcome of a negotiation eliminates the whole decision procedure in which one must constantly discriminate which details matter and which do not, since cooperative solution concepts only depend on what is available in each negotiation and therefore do not entail the evaluation of all the details of the negotiation process.

We should keep in mind that certain issues are important and directly affect the outcome of a negotiation. For example, the time factor[45] in negotiation is determined by various other factors such as, for example, the efficiency or impatience of both negotiators. But it also depends on factors such as culture, the forms of life of a society, whether the negotiation is to end in a short or long relationship, etc. A concept of cooperative settlement that does not take into account such details will be unlikely to predict an agreement. But how can we know under what circumstances a given cooperative solution concept[46] can be applied? The answer is not simple, and it depends on the value that detail assumes in any negotiation. If we obtain a cooperative solution concept that correctly predicts a negotiation by correctly evaluating the details significant in a given negotiation, then we will have found the correct procedure.

Nash's implicit proposition of this problem is complex to apply in bargaining, but it is an independent peculiarity of Nash's programme, so it seems appropriate to review his proposition. Instead of basing his argument on the cooperative solution concept proposed above, Nash chose to investigate bargaining from a non-cooperative point of view that would allow the prediction of the cooperative solution concept. Thus, whenever a negotiator uses a given cooperative solution concept to predict a bargaining outcome, Nash's programme requires us to ask the negotiator what he or she knows about the business practices usually applied in his or her professional discipline. We then develop a thought experiment that consists of devising a non-cooperative bargaining game that satisfies these requirements. If the result is suitable and solves the equilibrium selection problem, then we could predict the

44 As we will see below, this problem directly affects the principle of relevance. Cf: Grice (1989, 86–88).

45 Von Neumann and Morgenstern (1944) (2004, 724).

46 Nash (1953), 128–140 and Nash (1996, 34–46).

agreement that rational bargainers would reach if they were subject to the rules of the non-cooperative bargaining game. Undoubtedly, this agreement might or might not coincide with the prediction obtained using the cooperative solution concept. If the two should not coincide, the confidence of the applied negotiator in his cooperative solution concept would be unsustainable. Thus, the thought experiment would refute his proposal.

The application of untested cooperative solution concepts to predict negotiation outcomes has been heavily criticised. There is actually a close analogy between the non-cooperative bargaining games used to test cooperative solution concepts and the paradigms studied at any university. When constructing a model for a bargaining paradigm, a theorist does not aim to produce an exact replica of an actual negotiation of the purchase and sale of a financial product. In fact, there is usually only a rough idea of what such a financial transaction would look like. However, such a model will aim to align the properties involved in such a theoretical trade as closely as possible with usual financial product transactions as they take place in international markets. Once this goal is achieved, the simpler the model, the easier it will be to interpret the economic data. In fact, an experienced trader seeking to predict the outcome of a trade is unlikely to be adequately informed about the detailed structure of the trading procedure in use. Even negotiators themselves would find it difficult to enumerate all the twists and turns their negotiations might take. However, an experienced negotiator needs only a cursory notion of the product or service over which he or she is negotiating because he or she repeatedly possesses fuzzy information about many details when constructing a non-cooperative bargaining game for use in the Nash-developed programme. In fact, the less we know about the bargaining procedure used, the easier it is to refute the claim that a given cooperative solution concept will necessarily predict its outcome. The goal is always to start by constructing the simplest non-cooperative bargaining games that are consistent with the known facts, and the simpler the game we have to analyse, the less onerous the task of analysis will be. Of course, if we completely ignore the details that materially affect the outcome of the negotiation, we will be wasting our time when we attempt to employ Nash's programme, just as it is a waste of time to study abstract paradigms unless a student already has some clear idea about the characteristics of the financial market in which he or she would be negotiating.

It seems useful to list some of the elements that are crucial when rational traders negotiate. The outcome of a negotiation will depend on the object of the negotiation and on the negotiators' preferences about the different agreements they could reach. It will also depend on how the negotiators feel about the consequences of a disagreement. In both cases, the negotiators' attitudes towards risk or delay will be relevant. It is also strategically significant to know how disagreements can occur. For example, a breakdown in negotiations may occur because one of the negotiators prefers to opt for a different contract because he or she has been negotiating in parallel with alternative negotiators. Or it may be that the failure of a negotiation is due to force majeure beyond the control of the negotiators. Each factor has an impact on the final outcome of a negotiation. In addition, two notable considerations in the negotiation process can be enumerated. Nash explicitly recognises in the distinction between bargaining and negotiation that, first, it must be borne in mind that information about each negotiator's preferences is incomplete and second, that the preferences of the players are common knowledge.[47] The bargaining problem in the case of incomplete information is very complex. The bargaining model on which Nash based his system of axioms is known as the Nash demand game.[48] In this game, each negotiator simultaneously announces a demand. If both demands can be mutually satisfied, the demand of each is met. If not, both negotiators are confronted with the consequences of their disagreement. However, each pair of demands in the bargaining set of this game is a Nash equilibrium. Therefore, the problem of equilibrium selection must be addressed. Nash examined the dilemma making reference to the so-called "trembling hand perfect equilibrium" analysed by Richard Selten, according to which the risk of a wrong decision by the players must be taken into consideration.[49] When incorporating this uncertainty into the model, we obtain a unique Nash equilibrium outcome if the amount of uncertainty is sufficiently small. It is assumed that the equilibrium remains even with a low probability of error.

By constructing a demand game, Nash assumed the commitment possibilities available to the bargainers. From a theoretical point of view, a commitment is considered an irrevocable threat or promise, which cannot be retracted if subsequent events make one regret having been so rash. Sometimes, it is claimed that rational individuals are able to make a commitment in the present to perform an action in future. This commitment can be used as an instrument to achieve a bargaining outcome. Commitment must be approached with caution, however, when used as a bargaining instrument to achieve specific ends. The alternative to attributing unlimited commitment power to the negotiators involved in a demand game would be to recognise a higher-order state of affairs that arbitrates the negotiation in order to ensure that the rules are followed. However, such an assumption lacks force in business-as-usual.

5 Non-Cooperative Games

John von Neumann initiated a theory of two-person zero-sum games in his theory of partnership games published in 1928.[50] Some of his ideas had already been anticipated by Emile Borel,[51] but, as he later observed, without his minimax theorem, which proves that every game between two zero-sum players has a solution, game theory would not have been fruitful. The importance of von Neumann's achievements for economics seems to have gone unnoticed before Morgenstern persuaded von Neumann to collaborate in writing the book *Theory of Game and Economic*

47 Nash (1950, 155–162) and Nash (1996, 1–8).

48 Poza García et al. (2010, 26–34).

49 Selten (1965, 301–324, 667–689).

50 Von Neumann (1928, 295–320).

51 Borel (1924, 204–224).

Behavior.[52] This pioneering book aroused hopes that game theory would immediately transform economic theory into a predictive science of human behaviour. It is controversial to claim that integrating insights from game theory with data on interactive learning from psychological experiments will achieve this end. In the present volume, we want to close some gaps that game theory has left unanswered.

In fact, the part of Neumann and Morgenstern's book devoted to the theory of non-cooperative games dealt only with a special class of games: two-person zero-sum games.[53] All outcomes of such a game are zero for both players. Solving such games usually requires players to use mixed strategies. Such a strategy requires one player to play randomly with his pure strategies to mislead the opponent and prevent him from guessing which strategy he will be applying to the negotiation. However, the question arises, at what rate should each pure strategy be played? In their analysis of non-cooperative games, Neumann and Morgenstern propose that a player chooses these probabilities to ensure that his expected payoff in the game is never less than his security level, no matter what the opponent does. By security level, they mean the biggest advantage that can be guaranteed.

A player calculates his security level in a game by first calculating the minimum expected profit he could make using a mixed strategy. His security level would then be the maximum of the expected minimums. The recommendation for playing zero-sum games between two negotiators is called maximin. The maximin strategy is generally thought of as a decision-making model that asks for an alternative that can provide the best outcome in a worst-case scenario. The use of the maximin criterion is rational in two-person zero-sum games because the famous minimax theorem states that the combined security levels of both players add up to zero. Therefore, the minimax is an inverse method aimed to minimise the expected loss. To employ it, the negotiator starts from the assumption that he will be disadvantaged by his competitor's decision. In other words, he expects the worst-case scenario caused by the opponent's move. Neither negotiator will be able to obtain more than his expected security level. To sum up, due to the use of the maximin criterion, each player can guarantee his security level or improve it. As such, rational play in two-person zero-sum games must result in each player maintaining his security level.

However, Neumann and Morgenstern observed that two-player zero-sum games are not interesting in economics. Therefore, they asked what the appropriate generalisation is for games whose outcomes do not add up to zero. One school of thought, which persists to this day, simply takes the maximin criterion as a general principle of rational decision-making and applies it everywhere indiscriminately, which generates an unthinking mode of action, because if a player knows that the other player is rational and that it is rational to use the maximin criterion, then the first player will not use the maximin criterion. Therefore, he will play whichever strategy is the best response to the second player's maximin strategy. Unless the best response coincides with his own maximin strategy, as is the case with two-negotiator zero-sum games, a certain inconsistency arises when we consider the implications of the second player

[52] Von Neumann, Morgenstern (1944) (2004).

[53] Von Neumann, Morgenstern (1944) (2004, 169–219).

knowing that the first is a rational person who knows that the second is also rational. Nash later generalised the minimax theorem and considered it a minimum requirement for two strategies to provide a solution to a two-person game, requiring that each strategy be the best response to the other.[54] This pair of strategies, today called the Nash equilibrium, is basic to non-cooperative game theory. The proper extension simply states that, if a non-cooperative game has a solution, then it falls within the Nash equilibria. Suggestive games in economics often have many different Nash equilibria, the Nash demand game being a prominent example.[55] But if a game has a large number of Nash equilibria, which one should be selected as the solution of the game?

For a two-person zero-sum game, the answer to this equilibrium selection problem is irrelevant, because all Nash equilibria are equally satisfying. Nash further recognised that if rational negotiators observe the likely strategies of their opponents over time and gradually learn to adapt their own strategies to these to increase their payoffs, the process will eventually converge on a Nash equilibrium in bargaining. This means that negotiators find their way to equilibrium through trial-and-error learning. After all, negotiators do not normally negotiate haphazardly but tend to ensure a certain degree of certainty when negotiating. We will come back to this later.

6 Rien ne va Plus

We have recalled that John Forbes Nash specialised in game theory and negotiation processes.[56] He aimed to elucidate the typical structures of so-called non-cooperative games.[57] According to game theory, when two parties enter a bargaining process, a total of four possible outcomes arise. The optimal outcome reflects a situation in which both parties win equally—what the theory calls a "win–win game"—so that the maximum overall outcome is achieved. A win–win game is profitable because it is designed in such a way that all participants can benefit from the deal. In conflict resolution, the win–win strategy is a resolution process that makes all participants happy. The second and third possibility are situations in which the participants either "win-lose" or "lose-win" such that one party wins and the other loses. The result is a zero-sum game in which the total gain remains constant, as one party wins at the expense of the other. Both strategies belong to the field of non-cooperative games.[58] The fourth variant is the lose-lose game, in which both negotiating partners lose and no gains are made. In this variant, neither partner is willing to make concessions, so

[54] Nash (1951, 286–295) Nash (1996, 22–33).

[55] Nash (1950, 155–162), Nash (1996, 1–8).

[56] Nash (1996).

[57] Nash (1951, 286–295).

[58] Nash (1951), 286 ff.

Fig. 4 Payoff matrix with two negotiators choosing among two options

	Negotiator 2 chooses left	*Negotiator 2 chooses right*
Negotiator 1 chooses above	4, 3	-1, -1
Negotiator 1 chooses below	0, 0	3, 4

A game in normal form

the negotiation fails. In trade negotiations, this type of situation is avoided by using certain techniques to help resolve deadlocks.[59]

The game par excellence is the zero-sum game. The paradigm of this game may be illustrated by the following situation: two people who have jointly committed an armed bank robbery are cross-examined in different rooms. The stolen money cannot be recovered. The robbers were wearing masks, so they would not be recognised. The only offence with which they can be charged is illegal possession of weapons. When questioned separately, each person has the possibility to confess and implicate the other detainee or deny involvement in the robbery altogether. In this situation, the police can propose a plea bargain and, through the appropriate use of incentives, make both parties confess involvement in the robbery and consequently convict them both. This is known as the "prisoner's dilemma", and it is based on the assumption that maintaining cooperation is difficult because selfishness tends to lead the players to make self-harming decisions.[60] Often individuals opt for non-cooperation but this does not necessarily lead to mutual benefit (Fig. 4).[61]

The normal form of a negotiation (game) is represented by a matrix showing the players, their strategies and rewards, as depicted in the diagram above. There are two types of players: one selects an option on the horizontal axis, and the other one on the vertical axis. Each player has two strategies, specified by the number of rows and columns. The rewards are shown inside each axis. The first number represents Negotiator 1's reward in the row, the second Negotiator 2's reward in the column. If Negotiator 1 chooses up and Negotiator 2 chooses left, then their payoffs are 4 and 3, respectively. It is assumed that the players act simultaneously and do not know the other player's choice in advance. An extensive form of presentation is used when the players know about each other's strategy. In this case, a tree diagram is used, which shows the possible combinations of strategies associated with their respective payoff. Each vertex or node represents a point at which a player makes a decision.

[59] In the fifth chapter, we will analyse the mediation process and propose solutions to "lose-lose" games.

[60] Milinski has developed an optimal "tit for tat" strategy in game theory for the prisoner's dilemma: Milinski (1987), 433–435.

[61] In the next chapter, the prisoner's dilemma will be addressed from the perspective of rational choice.

The player is specified by a number next to the vertex. Lines starting from the vertex represent possible actions for each player. Rewards are specified in the leaves of the tree. Extensional games can also model simultaneous move games.

A *symmetric game* is a game in which a negotiator's payoffs for using a particular strategy depends only on the other negotiator's strategy. The game is symmetric if changing the identities of the players does not alter the payoffs of the strategies. Well-known examples of symmetric games are the "prisoner's dilemma", "the game of chicken"[62] and "the stag hunt".[63] In contrast, *asymmetric games* are those where there is no identical set of strategies for both players, as players develop their strategies in dependence on and in relation to the strategy of the respective other. In *zero-sum games,* the total benefit for all players in the game, in each combination of strategies, always adds up to zero. Thus, only one player benefits at the expense of all others. The negotiator gains exactly the amount that the opponent loses, which is everything that is at stake. In business, we are often confronted with non-zero-sum games, as some resolutions have net outcomes greater or less than zero. That is, one negotiator's gain does not necessarily correspond to another's loss. For example, a business contract ideally involves a positive-sum outcome, where each opponent ends up in a better position than he or she would have been in if no negotiation had taken place.

The "maximin" and "minimax" criteria state that each negotiator should minimise his maximum loss. The "maximin" criterion is set as follows: negotiator A determines that his minimum possible payoff is the largest possible payoff. Conversely, the "minimax" criterion determines that negotiator B chooses that the maximum payoff of A be as small as possible. Dominant strategy equilibria are fine when they appear in games, but unfortunately, this is rare in real life. Together, these strategies will constitute a Nash equilibrium if player A's choice is optimal given B's choice, and B's choice is optimal given A's choice. A Nash equilibrium can be interpreted as a pair of expectations about each negotiator's choice such that, when the other reveals his choice, neither will want to alter his action. In this situation, each negotiator knows and has adopted his best strategy, and all know the strategies of the other negotiators. Consequently, each negotiator gains nothing by changing his strategy if the other negotiator maintains his. Thus, each negotiator executes the best "move" he can. This generates a cooperative game between the negotiating parties, which in turn generates contractual compliance.

Cooperative game theory justifies contracts as long as they are intended to establish some stability in the business. In this case, two negotiators will bargain over the amount of investment desirable for a given contract. For example, Nash's proposed solution for bargaining demands that the investment be fair and efficient. Evidently, some negotiators may oppose the introduction of fairness criteria as their bargaining strategy will be oriented towards demanding more profit.

---

[62] The game of chicken describes a simulated conflict in which two participants risk their lives to the limit. The dilemma is best illustrated in a scene of the film "*Rebel Without a Cause*", in which two young men race their cars towards a cliff. The first one to jump out of his car loses. In such a dilemma, there are two possible outcomes, depending on the strategy the driver uses, e.g., drive straight ahead or turn or swerve before the precipice. Poundstone (1992, 195–212).

[63] Poundstone (1992, 101–129).

It should also be borne in mind that in any negotiation, simultaneous games may occur in which negotiators make offers and, at the same time, act in parallel, or in which negotiators are unaware of the other negotiator's previous moves. On the contrary, when subsequent negotiators have some knowledge of previous actions, we find sequential games. Knowledge does not necessarily have to be perfect; it need only consist of some information. For example, a negotiator may know that another negotiator did not carry out a certain action without knowing which of the other available actions he chose. This is a typical example of a game of imperfect information. On the contrary, a game of perfect information emerges if all negotiators know the moves previously made by all other negotiators. Thus, only sequential games can be games of perfect information, because in simultaneous games, not all negotiators know the actions being opted for by others. Perfect information is often confused with complete information. Complete information requires that each negotiator knows the others' strategies and payoffs but not necessarily their actions. In full information games, all negotiators possess the same relevant information. The interest in such a situation is usually not in deciding the best way to negotiate, but simply in deciding which negotiator has a winning strategy.

Clearly, negotiations in which the difficulty of finding an optimal strategy arises from the multiplicity of possible moves are called *combinatorial games*. Typical examples of games with a strong combinatorial character are those that include some imperfect or incomplete information. In the beginning, we introduced n-person games, i.e., those negotiations in which an arbitrary number of negotiators negotiate. In such negotiations, the parties negotiate repeatedly in their working life but consciously change strategies.

Negotiation considered from the point of view of game theory can also be approached in ethical terms. There are two antagonistic positions on this issue. On the one hand, there is the principle of grounding negotiation exclusively in terms of individual interest. This reflects Thomas Hobbes' theory of ethical behaviour among negotiators.[64] Since games like the prisoner's dilemma present an apparent conflict between ethics and individual interest, explaining why cooperation is necessary for individual interest is an important component of negotiation.

On the opposite side is Jean Jacques Rousseau's approach, according to which social games are rarely pure coordination games. To solve this "defect", he proposed modifying preferences and thus put forward the following procedure: if the general will could be fulfilled, then all individual wills would have to be brought into conformity with the general will.[65] Through this general strategy, a fundamental element on which the social contract is based was developed. This disparity of strategies has been outlined within game theory, as illustrated in the "stag hunt" dilemma, which describes a conflict between individual interests and social cooperation. To illustrate this dilemma, Rousseau described a situation in which two individuals go hunting. To hunt a stag, coordination between the two hunters is necessary. If they hunt one stag, they will have enough food for both of them for several weeks. However, if

[64] Hobbes (2007).

[65] The issue is addressed in Rousseau (2011), Book II, Chapter I. That sovereignty is inalienable.

Fig. 5 Stag hunt

		Hunter 2	
		Stag	Hare
Hunter 1	Stag	3, 3	0, 2
	Hare	2, 0	2, 2

they wish to satisfy their urgent needs, each hunter may want to hunt a hare instead, as this does not require coordination. As they hunt individually, each of them can eat their hares the same day. The following table shows the payoff that each possible combination produces (Fig. 5).

Rousseau pointed to the advantage that cooperation has for each hunter as it facilitates higher profit, even if they had to renounce individual autonomy. In other words, each hunter should separately choose to pursue the more ambitious and rewarding goal, giving up some autonomy in exchange for a higher joint gain. This situation is a model example and an analogy for social cooperation. The problem is that, although both hunters agree to hunt a stag together, there is always a risk that individual interests will overcome the general interest and one or both hunters will end up hunting a hare instead of a stag. Despite this risk, working together within a group of people with the same goals appears to be far more effective than pursuing goals individually.

So far, we have mainly focused on the combination of strategic moves and their possible outcomes. Now we will examine the communication that takes place among the participants. More specifically, we analyse the ways in which the participants communicate their preferences, intentions and goals to their counterparts. In the following section, we will study the language games that can contribute to the resolution of a conflict.[66]

7 Language Games

In game theory, we observe that negotiators talk to each other in order to gather information about the object of negotiation, exchange interests, develop strategies, coordinate goals, etc. This whole process is carried out through the use of language. Language is used both to negotiate and to pretend to negotiate. What is the difference between partners who really negotiate and those who only pretend to negotiate? Let us assume that "negotiating" consists of conducting the negotiation by fulfilling all the requirements that appear in a book on negotiation. However, someone could only create the impression with the other person of wanting to negotiate but does not really intend to do so. An ideal negotiator could have studied at a business school and could then show up for a meeting, perform flawlessly and all the time act as a serious negotiator. The question is how we can determine whether his actions are based on

[66] Wittgenstein (2021), Levinson (1997, 280f).

honest intent. We have assumed above that "negotiating" involves two requirements: that the negotiator has studied at a business school and has learned to negotiate. This implies that he knows the fundamental characteristics of negotiation, and that he is serious about it. But, we might ask, what does his school and his behaviour have to do with the business itself? We could, for example, observe that when he describes a product, he uses certain rhetorical statements. Additionally, he may use exaggerations and metaphors that leave his statements ambiguous. We notice that the negotiator's language becomes less exact, though no less pleasant.

Suppose someone on your own team points out that the other negotiator's choice of expression suggests that he or she might be cheating in the negotiation. One could respond to this objection by saying that a person who has studied at a prestigious business school would never be able to cheat. The terms on which a negotiation is conducted are generally interpreted according to the common intention of the parties, even if this differs from the literal meaning of their words.[67] If it is established that one party intended a term in the negotiation to have a particular meaning, and at the time of the negotiation the other party could not have been unaware of the first party's intention, that term must be interpreted in the sense of the first party's intention. Generally, if the meaning differs so much that the parties cannot agree on its specific meaning, then the terms should be interpreted according to the meaning imputed to them by reasonable persons of the same status as the negotiating parties who are also in like circumstances.

In interpreting the terms under which a negotiation is conducted, account should be taken of the circumstances in which the negotiations are conducted; the conduct of the parties, including after the negotiation has taken place; the nature and purpose of the negotiation; the interpretation already given by the parties to the negotiation; and the practices established between the negotiating parties. Account should be taken of the meaning commonly given to terms and expressions in a salient industry and the interpretation that similar clauses may have already received. Furthermore, account should be taken of the usage, good faith and fair dealing of the parties involved in the negotiation. It should be borne in mind that in case of doubt as to the meaning of a term that has not been negotiated by the parties, an interpretation of the clause contrary to the party that proposed it is generally preferred.[68]

It is generally assumed that terms are to be interpreted in the light of an entire negotiation. The negotiating parties are viewed as reasonable persons who intend the negotiation to be effective. Therefore, an interpretation that makes the terms of the negotiation lawful or effective is to be preferred to one that does not.[69] One of the

---

[67] Grice calls this domain the "conversational implicatures". Cf.: Grice (1989, 20–40).

[68] In negotiation, it is important that the contra proferentem rule applies, which means that in case of doubt, negotiated terms that have not been agreed individually will be interpreted preferably against the party that proposed them. In other words, the interpretation of ambiguous or obscure terms used in a negotiation should not favour the party that caused the obscurity.

[69] Any negotiation favourable to the lawfulness or effectiveness of the terms of the negotiation shall take precedence over interpretations that deny them. Therefore, if any term or expression expressed in the negotiation admits of different meanings, it must be understood in the most appropriate way for it to be effective.

most common problems in negotiation is due to unclear formulations or ambiguous statements used by the parties involved. International negotiations are usually drafted in more than one language in which ambiguities may arise due to translation errors. The parties can provide for a solution by stating that one version will be the authoritative version and will serve as the basis for the contract. If there is no such agreement, in the case of discrepancy, preference will be given to the version in which the agreement was originally drafted.

Suppose we all know that "negotiate" means to discuss an issue to reach an agreement for the purpose of conducting profitable business. If negotiators do not clearly determine the type of transactions to be conducted and do not assess potential benefits, they may be cheating. If a negotiator remains vague in his statements and does not give clear information about his preferences and goals, it can be assumed that he is not primarily interested in building a business relationship. He may be pursuing some other goal, whether he wants to learn more about the product details of the other company or whether he wants to test if the other is dependent on this business. This factor will be studied in the following pages to analyse the language games of negotiation.[70]

As is well known, all language is determined by polysemy. There is a general tendency to confuse the content of a term due to its referential variety. Due to the polysemic nature of most words, the language used in negotiation may be interpreted in different ways. First of all, this applies to the word "game" that may be misunderstood when used in the context of negotiations. If we compare the German or the Spanish translation with the English word "game", we notice that there is only one expression in German and Spanish, whereas the English language has the two different expressions "game" and "play". Both words have different connotations. Whereas "play" is synonymous with "recreational activity" or "amusement" carried out for pleasure and enjoyment, "game" is a structured form of play that follows certain predefined rules. In fact, many Spanish speakers confuse the ludic connotation of "play" with the regulated structures that all "games" follow. Wittgenstein has examined this difference and has underlined the rule-governed nature of games.[71] And this rule becomes evident by the use of an appropriate linguistic expression that follows our grammatical rules.

The expression "language game" used in negotiation refers to the use of specific linguistic procedures whose purpose is to carry out certain actions.[72] Language games are evidence of the fact that when a negotiator expresses himself in a specific language, he intends to carry out a specific activity, be it selling or buying something, acting in a certain way, acquiring a service, etc. The central element of language games is that negotiators use language in multiple ways. For example, in a given language game an expression may be used to refer to a particular object. Thus, when a speaker utters "Bring me the car", such an utterance can imply different actions

[70] Generally, the problem of sham bargaining is dealt with in game theory in the section on "bluffing". See: von Neumann and Morgenstern (1944) (2004, 186ff).

[71] Wittgenstein (2021, §23, §53).

[72] On the link between "language games" and speech acts see Levinson (1997) and Searle (1985).

depending on the situation and the salient communicative presuppositions. In one case the utterance is used to give orders; in another case, this could signal the answer to a question, such as 'What can I bring you?' or 'Shall I bring you the car or the truck?' or 'To whom shall I bring the car?'. Depending on the intention of the speaker or the circumstances, the phrase can take on different meanings.

All negotiations are intended to end with the drafting of a contract. Contracts are sometimes drafted in such a way that there may be divergences about the terms that have been drafted. The parties may offer different versions of the meaning of the terms. If nothing is stated in the contract itself and it is not possible to eliminate the divergences by other means—for example, by correcting obvious errors in the drafting—a reasonable solution has to be found by establishing the genuine meaning, as it is likely to better express the common intention of the parties. Generally, the original version is the one that should prevail. If there are still divergences in this respect, since the different versions may be regarded as indistinctly authoritative, then it seems advisable to respect the will of the parties by having recourse to the general rules of interpretation. In such conflicting cases, it is not possible to give preference to one version. It has to be decided which version corresponds best to the common intention of the parties or, if this cannot be established, what would be understood by reasonable negotiators in the context of the particular transaction. It is important to read this provision in the light of the contra proferentem rule if the original version was drafted by one of the parties, as this may have introduced the ambiguity causing this double interpretation.

The contra proferentem rule simply indicates that the interpretation of the negotiator who has introduced an ambiguous term should not prevail. In case of doubt as to the meaning of a clause that has not been individually negotiated, the clause negotiated by both parties and against the one who drafted it should be preferred. It is therefore assumed that the risk of any defect in the drafting of a contract is assumed de facto. The rule thus applies not only against the drafter but also against anyone who supplies predrafted or unilaterally drafted terms that have not been negotiated. One should account for whether the drafting has been carried out under the dominant influence of one of the parties. In this case, an interpretation contrary to the dominant party is preferred, since one negotiator is presumed to have exercised strong influence over the other. The contra proferentem rule applies only when the meaning of a term is doubtful. In many, if not most cases, the general rules applied in interpretation should make it possible to elucidate an unambiguous meaning.

The example developed above allows for the introduction of new cognitive biases. In addition to the equivocal interpretation in Spanish of "juego" understood as "play" or "game" indistinctly, we refer to multiple types of games: for example, board games, betting, we use it in sports, but also to refer to certain simulacra linked to armed conflict ("war games"). In all of these instances, the word "game" is used differently. A current example of polysemy is the redundant expression "no means no", which can be used as a reiteration of not wanting to do something; to express finality when negotiations break down; and more recently, it has been used as an expression against gender-based violence to indicate the gravity of violating non-consent in situations of sexual pressure. The meaning of words or an expression depends on the language game in which it is used by the speakers. Thus, a word or expression fixes its meaning through a negotiator's use of it within a given language game. A speaker may use the above expression to articulate outright refusal in negotiations, and it can also be used to warn someone that there has been no consent given within certain intimate relations. The expression could even be used as a code by members of a secret society or as an inside joke with a class of English speakers who are learning Spanish, both of which languages have the word "no".

It is important to note that one cannot limit the application of language games merely to the meaning of words or expressions. For example, the phrase "I want to buy a tonne of tomatoes" can mean several things. Regardless of the use we may associate with such a phrase, in the context of negotiation, the phrase itself is not clear since the other party cannot know whether it expresses an intention to purchase a certain product or is an indirect request or is meant to clarify that the tomatoes offered previously are not deemed suitable because of their poor quality and thus leverages the fact that the speaker holds some purchasing power given the quantity of tomatoes he or she wants to purchase. But the phrase can also mean that the company wants to buy a tonne of tomatoes *per month*, and that this is understood, or that they are looking for the cheapest price for this amount, or that the tomatoes are for salad or fried tomatoes or that they must be green. The full meaning of the phrase thus depends on its context of use. The phrase itself acquires its meaning when we identify the specific purpose to which the negotiator refers. It becomes meaningful once we set it in some context of negotiating. The phrase only has meaning when used to express something concrete. The expression "language game" is used in negotiation in a different way to refer to the following aspects, which are here summarily addressed:

First, fictitious examples of negotiating language use that are simpler than our own everyday language. These examples can be drawn from real dialogues, such as the dialogue between Chinese President Xi Jingping and Canadian Prime Minister Justin Trudeau in which they had a brief and tense interchange at the end of the G20 summit in Indonesia in 2022. The Chinese president told Trudeau through a translator: "Everything we talked about has been leaked to the newspaper. That is not appropriate. We don't act like that. If you are honest, we should communicate with mutual respect. If not, the consequences will be hard to predict". The Canadian president nodded as he listened to Xi's words and responded: "In Canada, we believe in open, free and frank dialogue. We will continue to look for ways to work together constructively, but there will be things we disagree on". Xi, whose gestures showed his discomfort, concluded the conversation by saying: "Let's create the conditions first". Afterwards, the two leaders shook hands and parted. Such language games between two leaders cannot be interpreted literally. It seemed from one side that the principle of confidentiality in communication had been violated and negotiations were going to have to be re-established by means of a new agreement as to the terms of conversation. It is necessary, in retrospect, to know whether all conversations between two leaders can be made public without the consent of one of the heads of state involved in the conversation. Furthermore, it must be known what kind of "confidentiality" is introduced into such talks and whether this principle plays a role in the dialogue between two countries' leaders.

Secondly, if someone does not know how to distinguish between information that is accessible and information that needs to be kept secret for security reasons, inexperienced negotiators should be trained to use the appropriate language of negotiation. This book will provide the necessary skills for mastering all the elements linked to the language games commonly used in negotiation.

Third, language games are used in a variety of specialised fields with their own grammars and relations to other language games. This is the case of language with special purposes.[73] Thus, a negotiator of a contract for vegetables will use an argot quite different from the vocabulary he would use in a contract for the sale of a ship. Specialised language is considered a system of linguistic varieties that manifests through very different language games according to the users and the communicative contexts in which the users are carrying out their negotiation. These language games used by negotiators are determined by the origin of the negotiators and the communicative situation in which they are negotiating. On the one hand, it should be borne in mind that the use of language depends on certain geographical, generational, educational, etc. varieties. On the other hand, the communicative situation of negotiators depends on the type of interlocutors, the medium of communication—oral, written, direct or indirect, etc.—the language used and who is using their mother tongue and who a foreign language, the intention of the parties in the negotiation and the issue to be addressed.

There is no agreement on the characterisation of the language games using special vocabularies vis-à-vis everyday language. In principle, it can be said that we are dealing with a subset of natural language, partially coinciding with the common language but used by specialists in a professional discipline. For some specialists, the language of negotiation consist of linguistic codes that share specific rules and units. Attempts have been made to analyse language games as lexical variants. Other studies have emphasised a particularly pragmatic point of view of these language games. What we can affirm is that there is a strong pragmatic component to how negotiators communicate with each other by means of a formal procedure. This allows us to distinguish between the use of language in general and the language games carried out in a special language between two negotiators, which we summarise as follows.

General language is used by speakers for conative, emotive, factual, etc. purposes. The subject matter is usually generic in nature since the discourse used by the speakers deals with common topics. In these cases, an informal communicative situation prevails. By contrast, the basic function of the language games used in negotiation is referential in nature. The topics addressed are specific, as the users are specialists working in a particular field. The communicative situation between negotiators is therefore formalised. The discourse on which communication is based is restricted to a particular domain.

Fourth, it should not be forgotten that any natural language is mastered when its user understands and uses particular language sets properly. These meanings are not separated from each other by clear boundaries but are mixed together. Language games are based on the following analogy: the rules of language are analogous to the

---

[73] Padilla Gálvez et al. (2003).

rules of games; thus, saying something in a language is analogous to making a move in a game. The analogy between a language and a game shows that words have a meaning that depends on the use made of them in the various and multiform activities of human life. Language games are not meant to suggest that there is something trivial about language, or that language is "just a game". On the contrary, the study of language games attempts to discover the rules that dominate communication within multiparty interactions.

The linguistic features of language games in negotiation are conciseness in form, precision in content, systematicity in structure as well as impersonality and objectivity in negotiation. At the phonological level, there may be interference of phonic elements among different languages. From a morphosyntactic point of view, there is a predominance of nominal elements (nouns and nominalisations) that express objectivity. Moreover, we observe a certain profusion of nominal adjacents (prepositional syntagms, specific adjectives, subordinate propositions, etc.) that express abstract meaning. The generalising article is used to achieve universality. In general, the present indicative with a timeless value dominates, lending universality and generality to the negotiating text. The past tense is used to refer to previous negotiation phases. The conditional, future and subjunctive are used to express hypotheses and conditions. In general, a negotiation begins with phrases containing gerunds and participle constructions, giving the discourse equanimity. The use of the infinitive with appellative value conveys a neutral and universal style. In the language games of negotiation, adverbs and other circumstantial complements occur that express clarity and precision. Impersonal and reflexive passive sentences are used to give impartiality. The analytical passive construction with inanimate subject without an agent complement is preferred. Explanatory adjective propositions are mainly used for precision and clarity. The presence of coordination should be emphasised, as it provides greater syntactic simplicity as it favours clarity. Clarifying clauses between commas, parentheses and dashes are also used for greater accuracy. Conjunctions with an equivalence value are used. Formulaic speech and fixed phrases help the interlocutors maintain procedural standards.

From a lexical-semantic point of view, language games used in negotiations contain abstract terms to enable universality. Definitions and technical terms are used for the sake of clarity. There is a certain predominance of structures containing enumerations. Neologisms are frequently found with the language's own procedures (by means of prefixes or suffixes) or with external procedures (borrowings, calques and acronyms). Finally, we note the use of non-linguistic elements such as numerical expressions, graphic elements, formalisations and symbols to refer to data or detailed descriptions.

The classic paradigm of a language game could be the negotiator's language: the language used in negotiation facilitates, in principle, communication between two negotiators A and B. For instance, negotiator A intends to purchase a product and negotiator B produces it. First, they have to clarify whether they agree on the characteristics of the product, and whether what B produces or offers is what A requires. Then all the terms and conditions are set out and the exchange is legally carried out. Throughout this negotiation process, a language is used that consists of specific words, certain procedures and the drafting of a contract in which the essential requirements, the acquisition of the product and its payment appear (assuming these are elements of a negotiation). This is done through the use of specific language sets for a particular business.

Let us analyse this process in a negotiation scenario in which two negotiators develop their different strategies. One wants to convince the other to buy a certain product. To do so, he will generate a discourse composed of specific language games. These language games consist of words and phrases that may be precise or ambiguous. To pursue a stated goal, the negotiators apply certain rules that shape their strategies. An opponent may use completely different language games in which the terms, statements and propositions will be more or less clear or confused. Within this communicative framework, what is most relevant is the way in which both interlocutors achieve a given end. This process is not as simple as one might think at first glance.

Several factors play a role in determining which language games are used by the parties involved in a negotiation. The cultural background and forms of life of the people involved in the negotiation form the framework within which the entire negotiation process takes place. If the partners are familiar with the business practices of the other negotiator(s), then this is a stable negotiating framework. It can become problematic if negotiating parties come from different cultures and do not know each other's business practices. A number of unforeseen situations may arise that require special attention and consideration. The crucial element in the negotiation process has to do with whether the partners correctly interpret each other's behaviour and language during the negotiation process.

Powell proposes three typologies of cultures with common peculiarities and calls them "*fact cultures*", "*people cultures*" and "*trust cultures*".[74] Each type is characterised by the use of certain business practices that are manifested in their respective language games. While negotiation partners of the so-called *fact cultures* focus their negotiation on content, facts and results achieved and keep small talk to a minimum; partners of the *people cultures* type tend to have a longer initiation phase with spontaneous changes in their agenda and a long decision-making phase. In contrast, mutual appreciation and reciprocal recognition play an important role in *trust cultures*, where negotiation objectives are communicated indirectly and decisions depend on mutual trust. In this cultural framework, certain language games emerge through which partners communicate their respective attitudes. Understanding the language game of one's partner requires interpreting it in relation to a given framework. For example, if a negotiator belonging to a people culture thinks "out loud" about an issue in the meeting and expresses a dissenting opinion or announces a new idea, this statement may fall on deaf ears or simply be ignored by a *fact cultures* partner. At the same time, the principle of never interrupting each other and always listening patiently, typical of the so-called *trust culture*, could be interpreted by a *people culture* negotiating partner as indifference and lack of interest.

---

[74] Powell (2012), 7–8.

To sum up, in negotiations interlocutors interact via language games to make decisions. A language game in negotiation can be studied through the rules used by each party in the negotiation process. In the negotiation process, both parties will try to reach an agreement by obeying certain rules, albeit in different ways. Language games enable us to analyse the rules used by each party in the negotiation. The purpose of the study of language games is the investigation of human behaviour in the face of decision-making. This process of decision-making and its form will be analysed in a subsequent chapter.

References

Arrow, Kenneth J. 1983. Economic equilibrium (1968). *Collected Papers of Kenneth J. Arrow*, vol. 2, *General Equilibrium* 107–132. Oxford: Basil Blackwell

Beth, Evert W. 1962. *Formal Methods: An Introduction to Symbolic Logic and to the Study of Effective Operations in Arithmetic and Logic* 40ff. Dordecht-Holland: D. Reidel Publishing Company.

Borel, Émile. 1924. Sur les jeux où interviennent le hasard et l'habilité des joueurs. In *Eléments de la théorie des Probabilités*, ed. J. Hermann, 204–224. Paris: Librairie Scientifique.

Bühler, Karl. 1965. *Sprachtheorie. Die Darstellungsfunktion der Sprache* 48ff. Stuttgart: Gustav Fischer Verlag.

Church, Alonzo. 1936. A note on the *Entscheidungsproblem. Journal of Symbolic Logic* 1: 40–41.

Gaffal, Margit. 2022b: La búsqueda de reglas en los juegos de lenguaje. Notas en los cuadernos de bolsillo. *Dókos,* 29–30, 213ff.

Gaffal, Margit. 2023: Ophelimity and the Paradox of the Stag or the Hare. *Dókos,* 31–32, 97–107.

Grice, Paul H. 1989. *Studies in the Way of Words* 20–40, 86–88. Cambridge Mass.: Harvard University Press

Hobbes, Thomas, 2007. In *Leviathan.* ed. Richard Tuck. Cambridge: Cambridge University Press.

Kamlah, Wilhelm, and Paul Lorenzen. 1973. *Logische Propädeutik. Vorschule des vernünftigen Redens* 211–213. Mannheim, Wien, Zürich: Bibliographisches Institut

Levinson, Stephen C. 1997. *Pragmatics* 227, 280. Cambridge: Cambridge University Press.

Lorenzen, Paul, and Kuno Lorenz. 1978. *Dialogische Logik.* Darmstadt. *Wissenschaftliche Buchgesellschaft* 4 (11f): 28.

Milinski, Manfred. 1987. TIT FOR TAT in sticklebacks in the evolution of cooperation. *Nature* 325: 433–435.

Myerson, Roger B. 1991. *Game Theory: Analysis of Conflict* 1ff. Cambridge Mass., Harvard University Press.

Nash, John F. 1950. The bargaining problem. *Econometrica* 18: 155–162.

Nash, John F. 1951. Non-cooperative games. *The Annals of Mathematics, 54*(2), 286–295. *Republished in:* Nash, 1996. 22–33.

Nash, John F. 1953. Two-person cooperative games. *Econometrica* 21: 128–140.

Nash, John F. 1996. *Essays on Game Theory* 1–8, 13–21, 23–32, 34, 48, 51–52. Cheltenham, Brookfield: Edward Elgar.

Padilla Gálvez, Jesús, (Ed.). 2016. *Action, Decision-Making and Forms of Life* 59ff. Boston, Berlin: Walter de Gruyter Verlag, 59ff

Padilla Gálvez, Jesús. 2021: *El mentiroso. Genealogía de una paradoja sobre verdad y autorreferencia* 226–232. Valencia: Tirant humanidades.

Padilla Gálvez, Jesús, and Margit Gaffal (Eds.). 2003. *Lengua de especialidad: Economía europea y derecho europeo/Language for Specific Purposes: European Economy and European Law/*

Fachsprache: Europäische Wirtschaft und europäisches Recht. Eds. Jesús Padilla Gálvez y Margit Gaffal. Cuenca: Ediciones UCLM.

Poundstone, William. 1992a. *Prisoner's Dilemma. John von Neumann, Game Theory, and the Puzzle of the Bomb* 101–129, 195–212. New York: Doubleday.

Powell, Mark. 2012. *International negotiations*, 7–8. Cambridge: Cambridge University Press.

Poza García, David J., Galán Ordax, and M. José, et al. 2010. Emergencia de normas espaciales de negociación en el juego de demanda de Nash. In *4th International Conference on Industrial Engineering and Industrial Management XIV Congreso de Ingeniería de Organización* 26–34. Donostia-San Sebastián.

Rousseau, Jean-Jacques. 2011. *El contrato social*, 18. Madrid: Editorial Gredos.

Schulte, Oliver, and James Delgrande. 2004. Representing von Neumann-Morgenstern games in the situation calculus. *Annals of Mathematics and Artificial Intelligence* 42: 73–101.

Searle, John R. 1969. *Speech acts.* Cambridge Mass.: Cambridge University Press.

Searle, John R. 1985. *Intentionality. An essay in the philosophy of mind.* Cambridge: Cambridge University Press.

Selten, Reinhard C. 1965. Spieltheoretische Behandlung eines Oligopolmodells mit Nachfrageträgheit, *Zeitschrift für die Gesammte Staatswissenschaft, 121*, 301–324, 667–689.

Turing, Alan, 1936–7: On Computable Numbers, With an Application to the *Entscheidungsproblem*, *Proceedings of the London Mathematical Society*, Series 2, 42, 230–265.

Von Neumann, John, 1928. Zur Theorie der Gesellschaftsspiele, *Mathematische Annalen, 100*, 1, 295–320.

Von Neumann, John, and Oskar Morgenstern, 1944. *Theory of Games and Economic Behaviour* 79–85, 94, 112, 154, 165–219, 221f, 238ff, 250f, 291f, 715–715, 724. Princeton: Princeton University Press

Wittgenstein, Ludwig, 2021. *Investigaciones Filosóficas.* 2ª edición. *Traducción del original alemán, introducción y notas de Jesús Padilla Gálvez* §23, §53, §143, §496. Madrid: Editorial Trotta.

Zermelo, Ernst. 1913. Über eine Anwendung der Mengenlehre auf die Theorie des Schachspiels. In Hobson, Ernest W, Augustus E. H. Love (Eds.), *Proceedings of the Fifth International Congress of Mathematicians* vol. II, 501–504. Cambridge: Cambidge University Press.

Chapter 3
Bargaining and Game Theory

1 Introduction

In the social sciences, bargaining is a type of negotiation in which the buyer and seller of a good or service negotiate the price or nature of a transaction. Its aim is to reach an agreement on the conditions under which the negotiation will take place and how the transaction is to be conducted. Although the most relevant aspect of bargaining is applied as an alternative pricing strategy to fixed prices, it can also include entering into credit or bulk purchase agreements, as well as serving as an important method of attracting clientele. The word "bargaining" can be applied to a wide range of situations in life.

Bargaining is directly linked to the forms of life that exist in societies. Thus, in Europe and the USA, it is generally limited to expensive or unique items and informal sales environments, such as flea markets. In other regions of the world, however, bargaining may be the norm, even for small commercial transactions. In Asia, bargaining for goods and services occurs everywhere, from street markets to hotels. There are exceptions, however, as food items qualify as basic necessities and are therefore considered insulting to bargain over as a tradable good.

In international negotiations, the bargaining process consists of two phases: first, generating value, and second, creating demand for this value. In some cultures, it is considered an offence if a party initiates the bargaining phase too early. In such contexts, it is important to create value first, which subsequently serves as the foundation of the business relationship. The reason that discontent can arise if this order is not followed is that in some cultures, especially in Asian countries, companies are normally interested in building a sustainable business relationship. In China, it is important to take time to build a relationship before starting to negotiate. In fact, failure to comply with this rule has often led to the termination of business relationships.

M. Gaffal and J. Padilla Gálvez, *Dynamics of Rational Negotiation*,
https://doi.org/10.1007/978-3-031-49051-4_3

The process of bargaining and its underlying dynamics may be illustrated with the help of game theory.[1] Bargaining games are played in certain situations in which two or more negotiators must reach an agreement on how to distribute a good, a service or a financial product. We are interested in applying bargaining games as a lens to situations like these. John von Neumann compared the bargaining process to what he calls "games" consisting of sequences of "moves" made by negotiators.[2] To illustrate this process identified by von Neumann, we make here certain general assumptions about the individual and group behaviour of negotiators acting in specific economic contexts. Based on these assumptions, we highlight some conflict situations that have a bearing on the resolution of the bargaining problem.[3] We start from the assumption that no hidden moves are made in bargaining. Therefore, we can draw a diagram of all possible bargaining sequences. As a rule, each player prefers to reach an agreement rather than to refrain. At the end of the bargaining process, we can draw a diagram outlining an actual bargaining negotiation. We assume that each negotiator wants to serve his or her interests. For the purpose of clarification, we divide the whole process into distinct parts, so that each stage of the process becomes visible.

According to the conventional view, the result of any negotiation is to reach an agreement between all parties involved. It is clear that one cannot yet derive a result from the individual steps taken by the negotiators. However, according to classical bargaining theory, each participant will choose one option among possible agreements, following the behaviour predicted by the rational choice model. More specifically, the preferences of each negotiator regarding the possible options can be represented by a function of the utility theorem developed by von Neumann and Morgenstern.[4] In the following sections, we will describe the possible steps in more detail.

2 Competition

Any negotiation that develops among two or more negotiators may result in a strictly competitive game. In such cases, as we briefly indicated in the previous chapter, the gains of one negotiator are always equal to the losses of the opposing negotiator. Such games are called zero-sum games.[5] They involve conflicting strategies in which the negotiators have completely opposite objectives and in which each player, for his part, seeks to obtain the maximum gain. Each negotiator seeks to maximise the losses of the respective other in order to obtain the inverse gain (Fig. 1).

---

[1] Von Neumann and Morgenstern (1944) (2004, 501, 512, 557f, 616), Nash (1950, 155–162), Nash (1996, 1–8).

[2] Von Neumann and Morgenstern (1944) (2004, 501, 512, 557, 616).

[3] A first attempt to present the different solutions proposed to bargaining can be found in Harsanyi (1956, 144–157).

[4] Von Neumann and Morgenstern (1944) (2004, 15f).

[5] Von Neumann and Morgenstern (1944) (2004), (85ff).

Fig. 1 Zero-sum game

Strategies chosen by A and B	B_1	B_2
A_1	2, -2	-2, 2
A_2	-2, 2	2, -2

Let us analyse these games using simple models in which few players participate and each player can choose one of two moves. The sum of all moves will always result in zero. We assume that there is no cooperation between the parties and that each player will choose the optimal strategy. If A_1 wins 2, B_1 loses 2, and if A_1 loses 2, then B_2 wins. If A_2 loses 2, then B_1 wins the same amount, and if A_2 wins it, then B_2 loses it. John von Neumann defined a rational form of decision-making for this type of game and called it the minimax principle.[6] The minimax principle consists in players developing a method of decision-making to minimise the maximum expected loss in games in which players aim to win the maximum. This requires perfect information of all the possible options.[7] Negotiators have perfect information if they know all the negotiating moves made in the bargaining process, so any decision they make will be based on this information. The principle is therefore based on a strategy aimed at minimising the least desirable scenario. The strategy has a counterpart in the inverse strategy, in which the minimum gains are maximised, called *maximin*.

Both strategies can coincide such that the negotiation is strictly determined. This situation is called the *saddle point*,[8] where a minimum and a maximum concur.[9] The saddle point is the point of agreement between the parties. Negotiators can on that basis conclude an agreement or a contract since, according to von Neumann, the saddle point presents a rational solution to any negotiation. The saddle point contains the following two properties: it is the largest of all row minima and at the same time the smallest of the column maxima. In his own words (Fig. 2):

> *The saddle point has the following two properties: it is* the highest of all the row minima and at the same time it is lowest of the column maxima. Thus, in Table 3a the row minima are respectively 1, 3, and 5, the last one being highest among them (*Maximum Minimorum*); on the other hand, the column maxima are respectively 5, 8, and 9 with 5 as the lowest (*Minimum Maximorum*). Hence the combination (A3, B1) yields both the highest row minimum and the lowest column maximun, and, therefore, constitutes a saddle point.[10]

The game situation can be described as follows: the minimums of the rows are, respectively, 1, 3 and 5, the last being the highest. This line is called "maximum

[6] Von Neumann and Morgenstern (1944) (2004, 89ff).

[7] Von Neumann and Morgenstern (1944) (2004), 112 f.

[8] In mathematics, a saddle point—or a "minimax" per von Neumann and Morgenstern (1944) (2004), 652—is a point on the surface of the graph of a function where the slopes (derivatives) in orthogonal directions are all zero (a critical point), but which is not a local extremum of the function.

[9] Von Neumann and Morgenstern (1944) (2004, 93f).

[10] Von Neumann and Morgenstern (1944) (2004, 652).

Fig. 2 Saddle point

Strategies chosen by A and B	B1	B2	B3
A1	2	8	1
A2	4	3	9
A3	5	6	7

minimorum". On the other hand, the maxima of the columns are, respectively, 5, 8 and 9, with 5 being the lowest, i.e., the "minimum maximorum". Therefore, the combination (A_3, B_1) produces both the highest minimum of the row and the lowest maximum of the column and therefore constitutes a saddle point. Thus, in the table, the saddle point is characterised by the strategy followed between A_3, B_1.

Vilfredo Pareto focuses on a different class of phenomena from those analysed so far.[11] The maximalist model proposed by von Neumann and Morgenstern presupposed that human beings strive to accumulate more and more goods. Their proposal is based on Hobbes' view, according to which people continuously seek to earn more money and maximise their profits. The asymmetric proposition outlined in the model of the zero-sum game helps explain these assumptions. The model of von Neumann and Morgenstern starts from the assumption that the amount of goods available to a player increases (or decreases) so that his new position will be more (or less) advantageous. But this model is not entirely correct. In fact, it is true that some quantities increase, such as for instance, the money earned in the game, but there are other measures that decrease, for example, the diminished time available with family for the one constantly aiming to maximise his profits. This raises the question of whether the consequences of this game are actually beneficial to the player.

On the contrary, Pareto is interested in studying the logical, repeated and numerous actions that people perform in order to secure the goods or commodity products that satisfy their tastes. To do so, a person must first be inclined to buy more than he needs to satisfy his tastes. However, he must take into account the price of the products.[12] As he cannot consume all the products, he will discard some of them to delimit his choice. In a second purchase, he will correct his mistake and, at least in part and little by little, will arrive at the desired number of items. At this point, he has reached an optimal state. His rectification has allowed him, through the logical application of a principle of refinement, to acquire what he likes in the right measure. In this model, the most desirable outcome for the player, but not the accumulation of profits per se, takes precedence. The model illustrates a momentum like the dynamic that is obtained in working environment. If one aims to improve one's output, one has to maximise one's profits, but this action is detrimental to one's quality of life since

[11] Pareto (1906, Chapter III), Pareto (2014, Chapter III).

[12] See footnote 11.

it prevents one from doing other things. However, this solution does not satisfy the negotiators' interests since none of them intends to maximise their interests by requiring the opposing negotiator to be willing to favour the first one. Evidently, the opposing negotiator sees no reason why he or she would be disadvantaged or lose everything just so the other party can take the full amount of the deal.

3 Cooperation

The most complex and least elaborate games are the opposite of zero-sum games, namely non-zero-sum negotiations. They stand for an equal interplay of interests and lead to a win–win situation in which all players gain something. They describe forms of cooperation between the negotiators. However, if cooperation is analysed from a rational point of view, it seems to work on a counterintuitive assumption, since all negotiators are supposed to collaborate to obtain results on equal terms, resulting in a win–win or a lose-lose outcome. Consequently, cooperation would nullify or at least reduce the principle of competitiveness between the parties. In fact, the principle of competitive advantage between negotiators would be limited if both negotiators aspired to cooperate with each other.[13] This incongruity needs to be clarified and will be discussed in more detail later. Let us first analyse cooperation in the light of game theory.

The view of negotiation as a non-zero-sum game is always accompanied by the possibility that each negotiator, when determining his strategy, assumes what may be called a "shadow" area in which negotiations are conducted with the sole purpose of defeating the other party by means of a zero-sum game. This shadow negotiation would undermine human interactions in such a way that cooperation would be prevented. We must not forget that shadow negotiations are underpinned by misleading and deceitful language games in which lies are told.[14] Under the appearance of cooperation, the negotiator's self-interest dominates under the surface. Let us take a closer look at this shadow negotiation and focus on the process and effect of one form of lie: namely, "bluffing".[15]

If the intended outcome of negotiation is a "win–win" result, then performing a given action, for instance, the purchase of a product, primarily serves to satisfy the negotiators' needs. Applied to the seller of the product, this means that he sells out of necessity. Hence, both parties receive a benefit: the buyer acquires the product at the agreed price, which will generate a certain profit, and the seller receives a benefit from the product transferred. It is worth mentioning that, for this to be fulfilled, the negotiation must be considered part of a long-term relationship between both

[13] The combination of "cooperation" and "competition" has given rise to the neologism "coopetition".

[14] Padilla Gálvez (2021).

[15] Von Neumann and Morgenstern (1944) (2004, 188f). Nash and Shapley (1953, 112–113, 115), Nash (1996, 17–18, 20). Bellman and Blackwell (1949, 600–605).

parties, since it will allow them to obtain mutual benefit. On the contrary, a "lose-lose" negotiation can still produce a benefit to third parties. The third party may simply be an intermediary, who is usually the only winner, since, regardless of the parties' benefits, he or she obtains the profits through commissions. Thus, in game theory, the mere introduction of an additional player can essentially alter the structure of the game.

Non-zero-sum bargaining is related to *integrative* or *interest-based bargaining*, in which the parties collaborate to find a win–win solution. This strategy focuses on developing mutually beneficial agreements based on the parties' shared interests. Such interests include each negotiator's needs, concerns and fears regarding their respective environments. All these encompass the underlying reasons why negotiators engage in a conflict. The aspect of *integration* refers to the parties' potential to combine their interests in a way that allows for a joint value to be generated. Integration can also lead to an expansion of the business. The potential for integration is only apparent in a negotiation if there are multiple issues involved. This is because the parties must be able to make concessions on the issues addressed as long as both parties are satisfied with the outcome.

We showed that the paradigm of non-zero-sum games is generally expounded by the prisoner's dilemma. The prisoner's dilemma involves two people who do not cooperate even if such non-cooperation works against the interests of both. The model was originally developed by Merrill M. Flood and Melvin Dresher and later formalised by Albert W. Tucker together with William Poundstone, who coined its name. The latter described it as follows:

> Two members of a criminal gang are arrested and imprisoned. Each prisoner is in solitary confinement with no means of speaking to or exchanging messages with the other. The police admit they don't have enough evidence to convict the pair on the principal charge. They plan to sentence both to two years in prison on a lesser charge. Simultaneously, the police offer each prisoner a Faustian bargain.[16]

The possible outcomes of such a deal can be summarised as follows: first, if A and B betray each other, each will serve 5 years in prison; second, if A betrays B but B remains silent, A will be released and B will serve 10 years in prison; third, if A remains silent but B betrays A, then A will serve 10 years in prison and B will be released; and, finally, if A and B remain silent, both will serve 2 years in prison for a lesser offence. A prisoner's dilemma point of view assumes that criminals who are offered a higher reward will betray their partners for the sake of individual benefit (Fig. 3).

If A and B accept the same offer, then loyalty to the partner is irrational, as both will receive between 5 and 10 years in prison depending on whether the treachery is one-sided. The prisoner's dilemma starts from the assumption that purely rational human behaviour is selfish[17] and that people will betray each other out of self-interest.

---

[16] Poundstone (1992, 8, 117).

[17] The prisoner's dilemma reinforces the thesis developed by Thomas Hobbes and postulated in *Leviathan*, in which he describes a human being who only follows his self-interest. See: Hobbes (2007, 2012).

Fig. 3 Prisoner's dilemma

	Silence of B	Treason of B
Silence of A	-2, -2	-10, 0
Treason of A	0, -10	-5, -5

Prisoner's dilemma

This implies that the only possible outcome for two purely rational prisoners is that they betray each other, even if mutual cooperation yields a greater reward. If we look at the table, we will see that betrayal is the dominant strategy for both players, which means that it is the player's best response given all the circumstances. This is because betrayal aligns with the principle of security, even if betrayal implies disloyal behaviour towards the other. However, such conduct is not treason from a legal point of view. This is because decisions made under collective rationality are not necessarily the same as those made individually.[18]

The prisoner's dilemma reveals that players act differently when their personal and collective interests prevail because certain cognitive biases are introduced in cooperative behaviour. These epistemic biases will be addressed systematically later. It seems appropriate to say a few words about the dilemma described above. The tendency towards cooperation has an impact on negotiation since it is important to delimit the negotiator's rationality and to know whether he is willing to trust a company or whether his negotiation is purely self-interested. Let us examine the experimental research carried out in this field. In one experiment, three forms of a game were adopted: a single game, an iterated game and an iterated game against a given player. Each game pursued different goals. The conclusion of this experiment was that if a negotiator wants to act rationally, then he should act as others expect him to act. Thus, the strategy to be followed must take into account the decisions of others and must include cooperation with other negotiators in order to succeed.

The prisoner's dilemma can be extended to achieve mutual cooperation between the parties. Its formalisation has shown that selfish defection can be rewarding. Suppose that in the game described above the parties were given a second chance. In this case, both negotiators wish to cooperate with each other again. However, given past experience, it is very likely that they will apply the tit-for-tat strategy.[19] Following this strategy, the losing negotiator in the prisoner's game will start by cooperating on the first move and thereafter change his strategy by doing what the opposing negotiator did on the previous move. This strategy is applied when a negotiator faces a potential predator, who usually applies the win-lose strategy. For this purpose, a

[18] This conclusion is confirmed by the "Tragedy of the Commons" dilemma. This dilemma describes a situation in which several players destroy a common resource, acting out of self-interest and independent of each other. The result is that community resources are vulnerable to excessive and irrational abuse by individuals in cases in which their personal interests take precedence.

[19] Milinski (1987, 433–435).

mirror system is introduced into the game. In many cases, it also requires two fictitious companions: a "fictitious negotiator" who cooperates and another who abstains. Generally, the fictitious negotiators behave according to the give-and-take strategy, thus corroborating the hypothesis that cooperation can evolve between egoists.

According to traditional theory, in the prisoner's dilemma, the usual rational choice in the face of such a dilemma is to give away something.[20] This solution has been called the maximising theory of rational deliberation.[21] The first assumption that follows is that the individual actions of both parties are rational if they are maximising. Negotiators try to get what they value most. The maximising actor argument posits that an interaction will be rational if its outcome is self-beneficial.[22] The goal of bargaining, after all, is to channel action in such a way as to achieve the objectives of one of the agents in accordance with their preferences. According to this view, the dominant choice is to maximise utility so that the outcome of the choice lies in individual gain.[23] This proposal focuses its actions on obtaining maximalist results. Actions are chosen following a strictly consequentialist rationality: the bargainer's actions must produce individual outcomes, and the reasons for acting are presumably reasons to promote such outcomes.

However, it has been objected that the outcome of maximisation theory is not optimal in all cases.[24] According to Pareto, a state is optimal if an alternative state cannot be produced that would improve the welfare of at least one participant without reducing the welfare of others.[25] If a change occurs that fulfils this condition, the new state is called a *Pareto improvement*. Thus, in the prisoner's dilemma, it is the case that if the offenders betray each other, both will have to spend the least amount of time in prison. This outcome is the opposite of the maximising thesis. In short, rational choice interpreted as maximising choice is not optimal in all cases described in the prisoner's dilemma. The rationality described in this proposition invites one to betray the other agent in order to maximise one's own benefit. From this point of view, the role of institutions and social practices is to mitigate undesirable outcomes. Thus, this situation can be improved by altering the net benefits of interactions or by the way negotiators evaluate their preferred actions, converging in an Pareto optimisation. This means that Pareto optimality produces a situation in which there is no action

[20] For a historical discussion of the prisoner's dilemma, see: Carroll (1987, 247–256), Howard (1988, 203–213), Kraines and Kraines (1989, 47–79), Lewis (1979, 235–240), Poundstone (1992), Kreps et al. (1982, 245–252).

[21] Gauthier (1975, 411–433). Cf.: Gauthier (2022, Chap. 1).

[22] The rationality of individual benefit is based on arguments that obstruct the cohesion of society and undermine social agreements. We are of the opinion that many of our reasons arise from our forms of life, on which social practices are based. In this sense, every negotiator assumes certain considerations that are rooted in these practices and embedded in the form of life in which he or she has been socialised.

[23] A separate issue, but also relevant in the context of negotiation, is that a negotiator who is driven only by his own interests and disregards the social practices in each society does not act morally.

[24] Gauthier himself raised the relevant objections against his own maximising proposal in favour of the Pareto optimisation: Gauthier, 2013, 191–204. Cf.: Gauthier (2022, 221f).

[25] Pareto (1906) (2014).

or allocation that makes one negotiator better off without making another negotiator worse off. But, for this situation to occur, a new theory of rational choice must be proposed.

A Pareto theory would require negotiators to seek an outcome that is optimal in the sense described above. It is evident that in the prisoner's dilemma, the optimal outcome per Pareto is superior to the outcome proposed by the participants' individual maximisations. The former presupposes that negotiators cooperate with each other. This view requires negotiators to achieve their goals—as measured by utility maximisation—by pursuing some kind of agreed outcome rather than each choosing the best option for him- or herself, i.e., taking the other's choices into account. Best-response theories insist that all cooperation must be based on a deeper level of non-cooperation, a view that seems to border on incoherence. People are concerned with achieving their own goals, and practical reason must surely be understood as conducive to this realisation.

The above objections to maximisation theory have opened up new trajectories in the research. One of the most suggestive alternatives has to do with an "optimising" approach, which allows us to interpret the prisoner's dilemma from different perspectives and facilitates a new reading. To this end, we shall analyse the interaction between the reasons that lead us to negotiate, the individual benefits gained from it and how such an interaction takes place within the framework of our social practices and our forms of life. The analysis of the interaction shows that both parties would do better not to seek their maximising response, but rather be guided by optimal mutual benefit. The goal of deliberation, after all, is to determine the action that would best realise the agents' objectives, given their preferences. Such a goal, if properly understood, is directed towards optimising, not maximising, and is in line with social practices and institutions that are based on cooperation and cohesion.

4 The Nash Equilibrium

How can a zero-sum game be considered a rational outcome if negotiators aim to receive the highest payoff to the necessary detriment of the other? The question is not easy to resolve since it confronts us with conflicting strategies. If one negotiator always plays the non-cooperative game, there will be consequences, and no one will be willing to do business with him. To prevent such a situation, let us assume that each negotiator knows the equilibrium strategies of the other negotiators. In such a situation, no negotiator wins by changing only his own strategy. If each negotiator has chosen a strategy and no negotiator can increase his own gain by altering his strategy, and thus, all negotiators keep their own strategy unchanged, then, John F. Nash observed, the current set of strategy choices generates a sort of equilibrium. In this case, the term "equilibrium" refers to an impasse, in that negotiators make no progress in the negotiation and do not find a solution to a situation in which, despite maximising individual gains, it is not possible to generate an incentive for altering strategy.

Suppose then that two negotiators N1 and N2 have gone to different business schools. The first has exclusively learned strategy E1 and the second has been taught only E2. Now both negotiators have a relationship and want to negotiate on a given product. Negotiator N1 will exclusively apply strategy E1 to the negotiation, which he knows and has learned. Likewise, negotiator N2 will only apply strategy E2, which he knows. In such a negotiation, N1 has no available strategy better than E1 for maximising his profit in response to N2 choosing E2, and N2 has no other strategy available that is better than E2 for maximising his profit if N1 always chooses strategy E1. This situation is interesting since both negotiators E1 and E2 over time will come to know their strategy and will continue to apply the strategy they have learned, thus producing the so-called *Nash equilibrium*. Now suppose that N1 and N2 negotiate with two new negotiators from abroad, N3 and N4, whose strategies are also different. In this case, the negotiation is extended and takes place between four negotiators <N1, N2, N3, N4>. The Nash equilibrium would be fulfilled if negotiator N1's strategy E1 is the best strategy with respect to <N2, N3, N4> and negotiator N2's E2 strategy is the best strategy vis-à-vis <N1, N3, N4> and so on.

Eventually, Nash defined the classical bargaining problem as a set of joint utility allocations, some of which correspond to what the players would get if they reached an agreement, and others that would represent what they would get if they did not.[26] A bargaining game for two players is defined as a pair <F, d> where F is the set of possible joint allocations of utility—that is, the possible agreements that two bargainers can reach—and d is the point of disagreement. The definition of a concrete bargaining solution as proposed by Nash is generally accepted. To this end, the axioms that this solution must satisfy are established. Some of the most frequent axioms used in the construction of bargaining solutions are efficiency, symmetry, independence from irrelevant alternatives, scalar invariance and monotonicity. The Nash bargaining solution is the bargaining solution that maximises the product of an agent's utilities in the bargaining set. However, the Nash bargaining solution only deals with the simplest bargaining structure. The most common pitfall of this approach is that it lacks dynamism.[27] For situations where the structure of the bargaining game is significant, a generalised game-theoretic approach is more useful. This can allow players' preferences regarding time and risk to be incorporated into the solution of bargaining games. It can also show that certain details may be relevant in bargaining and may influence its outcome. For example, the Nash bargaining solution to the prisoner's dilemma is different from the Nash equilibrium. In the next few pages, we will review the Nash games again.

The Nash bargaining game is based on cooperative bargaining.[28] Its solution is called the *Nash bargaining solution*. It is the only solution to a bargaining problem

---

[26] Nash (1950, 48f).

[27] Due to the lack of a dynamic factor, Nash's model does not provide an explanation for Pareto's results. Pareto defined that model as a situation in which it is impossible to benefit one person without harming another. Cf.: Pareto (1906) (2014) and Pareto (1920).

[28] Nash (1953, 128).

between two bargainers that satisfies the axioms of scale invariance, symmetry, efficiency and independence of irrelevant alternatives. As we describe repeatedly in this book, the Nash bargaining game is a simple game involving two players that are used to model bargaining interactions. In the Nash bargaining game, two players demand some good and bargain over the matter to be negotiated.[29] If the total amount requested by the players is less than the amount available, both players get their claims. If the total amount requested is greater than that available, then neither player satisfies his demand.

Nash presented a non-cooperative demand game with two players who are not sure which pairs of payoffs are feasible. In the end, as uncertainty disappears, the equilibrium payoffs converge to those predicted by Nash's bargaining solution.[30] This calls for a brief explanation of the equilibrium analysis. Suppose that the strategies are represented in a Nash game by a pair (x, y), where x and y are selected from the interval $[d, z]$, where d determines the outcome of the disagreement and z the total quantity of the good. If $x + y$ is equal to or less than z, the first player receives x and the second player receives y. Otherwise, both receive d. There are many Nash equilibria in such a game, namely any x and y such that $x + y = z$ is a Nash equilibrium. If one of the players increases his demand, then both receive nothing. If either player reduces his demand, he will receive less than if he had demanded x or y. There is also a Nash equilibrium in which both players demand all of the good. In this case, both players receive nothing. However, neither of them can increase his return by unilaterally changing his strategy. Obviously, these mathematical realities require a solution in the real world.

In the secondary literature, several solutions have been proposed based on slightly different assumptions about the desired properties for the final agreement point of Nash bargaining.[31] For this purpose, certain axioms were introduced, but Nash showed that the solutions satisfying these axioms are exactly the points at which the players act as if they are seeking to maximise their profits provided that they stop bargaining with the other player. In this case, each player obtains his utility from non-cooperation and a share of the profits from cooperation.[32] Another solution is to carry out a sort of egalitarian negotiation, granting an equal profit to both parties. This is intended to maximise the minimum gain between the players.

29 Nash (1950, 155f).

30 Nash (1953, 129f).

31 Kalai (1977, 1623–1630).

32 P. Champsaur studied the solution concept taken from the theory of cooperative games based on a full information hypothesis. When imperfect competition is involved, cooperative behaviour with complete information may seem a limited framework. However, the solutions provided by the theory of cooperative games are the only ones that give us useful insights into what conditions would allow the emergence of perfect competition. Cf.: Champsaur (1975, 394–417).

5 Mixed Strategies

Obviously, pure strategies are rare, and in many games, there are no saddle points as described above, so there is no way for two players to converge at any given equilibrium. Naturally, negotiators do not learn a single strategy but have a range of strategies that they deploy in a single negotiation. Nor do they reveal their strategy during the game, and if they believe that the other party has become aware of their strategy, they develop "smokescreen" tactics to hide it by resorting to bluffing. Another factor to take into account in game theory is that the parties can invent the rules and thus apply new strategies that in turn generate gains. Mixed strategies are carried out by randomising of pure strategies. This procedure is carried out by assigning to each pure strategy a probability indicating the frequency with which each strategy will be played. This procedure is carried out by the game of matching pennies—which von Neumann and Morgenstern characterised in their book as the "Game of Matching Pennies"—or the children's game of "rock, paper, scissors".[33] They also addressed the Sherlock Holmes dilemma.[34] The children's game is a good example of mixed strategies, so it is worth describing.

The players count down together "rock, paper or scissors!" At the end of the count, they all show one of their hands at the same time, so that the element that each one has chosen can be seen: stone, a clenched fist; paper, all fingers extended; and scissors, index and middle fingers extended and forming a "V". The objective is to defeat the opponents by selecting the weapon that wins according to the following rules: stone crushes scissors, scissors cuts paper, paper wraps stone. In case of a tie, where both have chosen the same element, the game is played again. The game is repeated until one of the players wins by a specific number of points previously agreed upon. The best way for one of the players to win the game is to play randomly. This is called "mixed strategy", as opposed to "pure strategy", which consists of always betting on one element.

If a random strategy is designed from scratch, players can generate an equilibrium that reinforces itself during the game. Thus, von Neumann and Morgenstern demonstrated mathematically that if a player plays randomly the games of "matching pennies" or "rock, paper, scissors", he has the same probability of winning and losing as long as the opponent plays with a pure strategy or chooses it randomly. The average profit for the player who plays randomly is zero. For this reason, the random strategy has the maximum minimum. Now, if the second player starts the game, then it would result in a minimum maximum gain for the player applying the random strategy. Both players playing the randomised strategy would result in an ultimate equilibrium between both parties.

A most disturbing issue is the Sherlock Holmes dilemma, in which scores are assigned to calculate a mixed strategy.[35] Sherlock Holmes has been investigating Professor Moriarty and his accomplices for months and is on the verge of catching

[33] Von Neumann and Morgenstern (1944) (2004, 111).

[34] Von Neumann and Morgenstern (1944) (2004, 177f).

[35] Doyle (1893) (1993, 583–589).

Fig. 4 Mixed strategy

	1	2
1	100	0
2	-50	100

them and bringing them to justice. However, Moriarty has organised a highly sophisticated crime syndicate. Holmes believes that if he can defeat Moriarty, it will be the greatest achievement of his career. However, Moriarty is willing to thwart Holmes' plans, as he is his intellectual counterpart. Sherlock Holmes wishes to go from London to Dover and then embark for the Continent, as he longs to escape from Professor Moriarty. Watson arrives at the station and boards a first-class carriage reserved for Holmes, in which he finds Holmes disguised as an Italian priest. As the train leaves Victoria Station, Holmes discovers Moriarty on the platform, trying to make the train stop. With the train moving, he notices Moriarty on the platform. Sherlock surmises that Moriarty might take a special train and overtake him to murder him. Sherlock can get off the train at Canterbury or Dover.[36] The crucial question to be solved following game theory is: Where will Moriarty get off? (Fig. 4).

Von Neumann and Morgenstern depicted the matter by means of a table.[37] The fields (1,1) and (2,2) correspond to Professor Moriarty capturing Sherlock Holmes and murdering him, so the higher coefficient is assigned. The field (2,1) indicates that Sherlock Holmes escapes to Dover as Professor Moriarty will get off the train in Canterbury. The field (1,2) means that Sherlock Holmes escapes from Professor Moriarty, as he will get off at Canterbury as Moriarty continues to Dover. However, Sherlock is unable to embark to the Continent and finally escape. In "*The final Problem*", Sir Arthur Conan Doyle opts for the last play—what he calls the "*coup de maître*"[38]—preferring to solve the unknown by applying the mixed strategy and thereby saving the famous detective even though the odds in that game fall in Professor Moriarty's favour. In fact, according to the calculations of von Neumann and Morgenstern, the probability of Sherlock Holmes dying is 48%.[39] If we approach Conan Doyle's story from the point of view of game theory, the author of the novel has probably premeditatedly deceived the readers, bluffing in that he made a false move, or a strong bet deceive the readers. Evidently, if he had not lied, Sherlock Holmes would have died at the hands of Professor Moriarty, so von Neumann and Morgenstern conclude that when Doyle performs the so-called *coup de maître*, the rational gamble does nothing but bluff in this passage in order to save his protagonist from a death foretold.

---

[36] Doyle (1893) (1993, 584).

[37] Morgenstern (1935, 337–357). von Neumann and Morgenstern (1944) (2004, 714).

[38] Doyle (1893) (1993, 585).

[39] See the footnote in the work of von Neumann and Morgenstern, where the mathematical proofs of the argument informally outlined above are presented and their calculations in this respect are set out: von Neumann and Morgenstern (1944) (2004, 178, note 1).

The economic conception of bargaining rationality is based on maximising the bargainer's utility. Maximisation is considered as a decision-making rule for maximising the gains in the best-case scenario in which a negotiation takes place. Its formulation was carried out in the theory of multiplayer zero-sum games, covering both cases in which players make alternative moves and those in which they make simultaneous moves. It can also be extended to more complex games and to decision-making in general in the presence of uncertainty. The negotiator's utility maximisation thesis, abbreviated ($\text{Thesis}_{\text{max}}$), is defined as follows:

$\text{Thesis}_{\text{max}}$ Every negotiator must satisfy his preferences against the predilections of the other negotiator.

The ($\text{Thesis}_{\text{max}}$) simply states that the negotiator regards society as a competitive system. This means that every negotiator assumes that social and economic relations are based on the development of advantages over competitors. Therefore, the goal pursued by every negotiator is to obtain a prominent position in his or her environment. Competitiveness is considered a skill that enables the development of resources or technologies and is based on the application of certain attributes that make the negotiator who possesses it superior. Obviously, the concept of "advantage" is relative since it compares the performance of a negotiator or organisation to that of others.

The ($\text{Thesis}_{\text{max}}$) produces mainly zero-sum games. The deals a negotiator makes are generally focused exclusively on realising his own interests and achieving his own gains. The economic conception of maximalist rationality links a negotiator's reasons for acting directly to his or her preferences of outcome. The negotiator evaluates actions one by one, taking into account exclusively his probable realisation of the outcomes as chosen exclusively by him. His action focuses exclusively on the outcomes of his strategy. Therefore, when he alters his preferences in the context of the negotiation, this change does not take place through the link between preference and action, but rather he appreciates the relevance of the sequences of actions by adjusting them as closely as possible to the plans that emerge from a zero-sum game.

However, the ($\text{Thesis}_{\text{max}}$) is not rational. It assumes that profit maximisation always takes place in relation to a competitor and never considers that competitive shortcomings may be due to errors within the system itself. Therefore, the main thesis is that ($\text{Thesis}_{\text{max}}$) can only improve performance in opposition to an external bargainer and at the expense of this bargainer losing or having its own gains offset. Assume that rational action takes place through complex processes in which the evaluation of possible outcomes arising from changes in expected or experienced preferences changes. Thus, a change of perspective brings different preferences into play. As one carries out a plan, one may find that one's evaluations of possible outcomes change, as one's preferences change, or as different preferences come into play in one's choices. Let us look at the process by way of an example:

Suppose a leading enterprise produces a particular product or service and its production system is obsolete because it is labour intensive. However, the employees of this enterprise guarantee with their vote a social and political status quo in the region and in the state, forestalling the appearance of radical parties. However, elsewhere, another smaller firm has emerged that produces a similar but more competitive and lower-priced product or service. The larger firm applies a preference-oriented maximising rationality and acquires the smaller firm in another country whose product or service is more advantageous. In principle, it is thought that the leading firm wants to acquire the new technology and improve its market position by diversifying. However, its maximising strategy is completely different: it closes the factory in order to secure jobs at home in order to maintain the social and political status quo. In this situation it can be seen how the (Thesis$_{max}$) is not always applied to competitively improving the product or service, but often endures irrational elements and is unable to solve the problems of internal adjustment within a company itself. In short, the working hypothesis of (Thesis$_{max}$) makes it impossible to improve internal strategies in favour of a competitive system oriented towards diminishing exclusively external profits.

This example easily shows that the (maximising thesis) and the zero-sum strategy make an internal improvement of the rational model impossible and block any progress, since its strategies are simply oriented towards satisfying its preferences by way of winning over its opponent. Obviously, the maximising thesis has drawbacks, since its orientation is based on rivalry with an outsider and will never allow, or even block, any internal improvement. In order to improve rational structures, we are faced with a paradigm shift: concurrence must be replaced with or at least supplemented by a cooperative model aimed at optimising the internal system. The negotiator's utility optimisation thesis—abbreviated, (Thesis$_{optimus}$)—is defined as follows:

Thesis$_{optimus}$ A state is optimal if an alternative state cannot be produced that improves the welfare of at least one negotiator without reducing the welfare of other negotiators.

Clearly, if two negotiators are not willing to accept a minimum consensus by cooperating with each other, then any reason introduced into a negotiation for adopting an optimisation is doomed to fail, as such reasons will either be false or merely a bluff. Even if we accept that negotiators are trying to achieve an optimal outcome whose benefits are acceptable to all, and that negotiators regard cooperation as directly rational, their ends may be different from their original intentions. The problem is that negotiators do not indicate when they negotiate that they will exclude the possibility of cooperation and mutual prosperity in future.

This outcome can be described in game theory by means of *matrix games*. Matrix games are simplified decision situations that can be represented in a matrix (in a table). Suppose there are two negotiators who can decide in two alternative ways. One negotiator is assigned the row of the table (row negotiator), the other the columns of the table (column negotiator). The number of rows and columns reflects the decision options of each negotiator. The results of the game are evaluated by means of pay-out values (profit, outcome). The pay-out amounts are entered in the corresponding table fields. The convention is to first enter the outcome of the row trader in the table field and then, separated by a comma, the outcome of the column trader (Fig. 5).

		Column negotiator (S)	
		A_1	A_2
Row negotiator (Z)	A_1	$E(Z)_{1\text{-}1}, E(S)_{1\text{-}1}$	$E(Z)_{1\text{-}2}, E(S)_{1\text{-}2}$
	A_2	$E(Z)_{2\text{-}1}, E(S)_{2\text{-}1}$	$E(Z)_{2\text{-}2}, E(S)_{2\text{-}2}$

Fig. 5 Pareto dominant/Nash equilibrium

The alternatives are denoted A_1 and A_2. $E(Z)$ and $E(S)$ are the results of the row and column negotiator, respectively. The outcomes are evaluated according to the following criteria. Efficiency is based on the Pareto criterion, i.e., the outcome of a negotiator that cannot be improved without making the other negotiator worse off. Furthermore, the Nash equilibrium is valid, according to which there is no incentive for the negotiators to abandon the negotiation. Consequently, it must also be considered that an outcome will be Pareto dominant if it exceeds the Nash equilibrium. If we analyse the game, we obtain the following result: the Pareto optimum will be the square with the highest payoff; the Nash equilibrium is the square chosen by both players. Depending on the occurrence of Nash equilibria and their coincidence with the Pareto optimum, games are divided into harmonious or conflictual ones.

Both models inform us to some extent about the processes involved in negotiations. However, the impossibility of homogenising negotiations becomes apparent in the way non-logical actions are analysed. In these cases, the formal models present notable difficulties. And in the case of the above models of negotiation, even if they are limited to certain games and their structure, the difficulties are palpable, as the case of Holmes has shown. Indeed, the sudden and unexpected way in which the bluff appears in the negotiation generates problems for the models described. For this reason, we are forced to introduce into our proposal, in addition to the analytical models, certain categories of social phenomena that appear in negotiation and substantially alter its ends. Let us briefly analyse these phenomena by means of a few simple examples.

6 The Functions of Bluffing and Lying in Negotiation

The case of Sherlock Holmes ends in a paradox since Doyle introduced a new element into the story. In fact, he presented a new element of game theory: bluffing and lying.[40] The bluff described in "*The Final Problem*"[41] is based on providing a false solution to a puzzle that should be solved rationally. In short, the bluff is based on a lie

[40] Padilla Gálvez (2021).

[41] Doyle (1893) (1993, 576ff).

that provokes an irrational behavioural change in Holmes. As we will see in other chapters, bluffing and lying play an important role in business and, therefore, in game theory. Bluffing is a strategy like that of the smokescreen, since the first thing that is intended is that the other player or negotiator stops thinking and the rational process be supplanted by a guessing procedure. What the other party must not understand is the rule on which his strategy is based. Most of the time, the negotiator has only learned one strategy and is therefore constantly forced to act randomly and lose small sums in order to get the maximum sum through his zero-sum game. Therefore, players choose to bluff and lie prudently so that their behaviour is not predictable. In the following pages, we will present some important elements of this for game theory.

In economics, bluffing has been explained as rational equilibrium behaviour in games with information asymmetries. This asymmetric information usually appears in bargaining relationships in which both negotiators possess "imperfect information"[42] about the other party. The dilemmas that arise in this respect are of a social order, since each player may share the same preferences in the negotiation, but the order of preference nevertheless does not coincide. Thus, both negotiators may intend to make a common bargain, but for one of the negotiators, the interest of the company prevails while for the other personal interest prevails. This also occurs in an inverse form. For example, let us consider the problem of bargaining deadlock, which is a central problem in contract theory when information is incomplete. The deadlock problem describes a situation in which two negotiators could work efficiently by cooperating in bargaining, but refrain from doing so for fear of giving the other party greater bargaining power and thus reducing their own gains. In both examples, we observe that the negotiators do not agree in their bargaining, as each gives a different meaning to utility—in terms of individual and collective gain—and the tangible rewards resulting from bargaining, which for one should translate into gain for the company and for the other into personal gain. Therefore, the terms "profit" and "gain" are not equal in the negotiation process.

To understand the process in which the deadlock is established, let us assume that there are two investors who invest a certain amount of capital.[43] Let us assume the following scenario: Investor A can make an investment today and Investor B could buy it and pay him the profit on this investment. If investor A rejects the offer, he will only be able to obtain on his own a fraction ($x < 1$) of these returns. Suppose that investor A possesses insider information about x. Then investor A could make a large investment even if it were a weak investment. The reason would be that a large investment could induce investor B to believe that investor A is making a major investment—which would entail recognising that x is relevant—so investor B is likely to make a generous offer that would exceed the fraction of the returns. In this case, bluffing is a profitable strategy for investor A, as it would afford him a greater return on investment.

[42] Von Neumann and Morgenstern (1944) (2004, 164).

[43] See the mathematical solution to a case in which two persons perform a game in which a bluff is applied in: Bellman and Blackwell (1949, 600–605).

Evidently, bluffing belongs to the strategic sphere and is therefore used systematically in various games. It is a false move that pursues different purposes: first, to divert the opponent from a correct strategy; second, it is applied to alter a permanent optimal option; third, it is used as a defence strategy; fourth, it is also applied as an offensive strategy.[44] Thus, we observe that bluffing is used in opposing strategies to induce the other party to change his behaviour and to derive some profit from this change. Usually, a distinction is made between a low- or high-risk bet. If a bluff is applied, one can lose or win in either case. We have indicated that in semantic terms, bluffing is equivalent to lying directly or inductively making the other party think something that is not accurate. This form of lying is intended to make the other party alter its behaviour by means of diversion, alteration, incongruous defence, or inconvenient attack. It thus seeks to divert the other party from the right strategy or decision, or to embolden him and attack where there is no gain, or to act cautiously when he could obtain a great return. Thus, the objective of a bluff is to induce at least one opponent to retreat or to improve conditions.

Bluffing produces a form of consensual "lying" within the framework of the game. In fact, the bluffing liar in the game remains monologically vague and all his information is passed on through non-verbal language. His main objective is to obtain information from the other party. In his strategy, there are neither facts nor truths; he has no moral scruples or empathy but pursues a sure gain by substantially applying the zero-sum game to achieve maximum individual gain. For all these reasons, language—both its use and its omission—is an effective procedure for controlling others. The behavioural pattern of liars who apply bluffing as an instrument is carried out by means of a mixed strategy of procedures in which seduction processes are integrated to dominate the game by manipulating and exploiting prejudices. To achieve their ends, bluffers seek to undermine the principles on which human relations are based and to play by their scruples. It should be borne in mind that the use of bluffing in the game has long-term effects, as negotiators who abuse this procedure induce the termination of relations between parties. For all these reasons, bluffing is mainly used in business and relationships characterised by brevity, where there is no personal relationship between negotiators.

In theory, bluffing behaviour is different depending on whether it is applied in a high-stake or low-stake scenario. The solutions for each of these alternatives were mathematically systematised and described in detail by von Neumann and Morgenstern.[45] To summarise, in negotiation, a pure bluff is considered a bet with lower exit conditions that have little or no chance of improving. A negotiator who uses a bluff believes that he can win if all opponents fold. The pure bluff strategy will be profitable for a player if the probability of being called by an opponent is less than the odds of the gain to be achieved by bluffing. In games with multiple betting rounds, bluffing in one round with an inferior or matched hand that could improve in a later round is called an inferior bluff, or even a semi-bluff. A player who makes this move can win in two different ways: on the one hand, if he succeeds in inducing the other

[44] Von Neumann and Morgenstern (1944) (2004, 205f).

[45] Von Neumann and Morgenstern, (1944) (2004, 216–219).

players to fold immediately; on the other, if he comes upon some new condition that improves his game. This type of play occurs in some cases where a player may find himself in a tie but with high enough odds to win the game.

It should be noted that a bluff introduces an element of risk into the game.[46] A bluff can be a great success or a great failure. Risk is a measure of magnitude that quantifies the damage in the face of a given gamble that can even frustrate a business. Risk is measured by assuming a certain vulnerability. When we refer to risk, we mean the occurrence of potential damage or harm to individuals and legal entities. In fact, the greater the vulnerability, the greater the risk; but the more feasible the harm or damage,[47] the greater the danger.[48] Therefore, the risk created by a bluff in a negotiation refers only to the possibility of harm under certain circumstances. On the contrary, when we refer to danger, we only refer to the probability of harm under those circumstances. For example, from the point of view of the risk of harm, the greater the lack of knowledge of a market in which trading begins, the greater the risk of harm to its traders, while the greater the imprudence in trading, the greater the danger of losing, for example, financing or making an investment in which all the capital or investment made is lost.

References

Bellman, Richard, and David Blackwell. 1949. Some two-person games involving bluffing. *Proceedings of the National Academy of Sciences* 35 (10): 600–605.

Carroll, John W. 1987. Indefinite terminating points and the iterated prisoner's dilemma. *Theory and Decision* 22: 247–256.

Champsaur, Paul. 1975. Cooperation vs competition. *Journal of Economic Theory* 11: 394–417.

Doyle, Arthur Conan. 1993. The final problem (1893). In *The Great Stories of Sherlock Holmes* 576ff, 583–589. London: Magpie Books.

Gauthier, David. 1975. Reason and maximization Canadian. *Journal of Philosophy, IV* 3: 411–433.

Gauthier, David. 2013. Achieving Pareto-optimality: Invisible hands social contracts, and rational deliberation. *Rationality, Markets and Morals* 4: 191–204.

Gauthier, David. 2022. In *Rational deliberation: selected writings*. ed. by S. Dimock, Cl. Finkelstein, and Ch. Morris 221f. Oxford: Oxford University Press.

---

46 In game theory, the choice of players in situations of risk and uncertainty can generate some confusion, though the two terms are sometimes interchangeable, as Frank H. Knight pointed out. We owe to his work a fundamental distinction. Generally, in a situation involving risk, the outcome is unknown, but the distribution or probability of its occurrence is known. These probabilities can be objectively or subjectively specified, which means that players will form their expectations with respect to these probabilities. Therefore, risk is said to be a "known-unknown" but uncertainty is an "unknown-unknown", since a player cannot assign probabilities to each outcome. Cf.: Knight (2002).

47 The terms "harm" and "damage" are not synonymous, and they need to be distinguished within game theory. We speak of harm if a negotiator is deprived of any lawful gain, which should have been obtained with the fulfilment of the obligation generated in the game. Damage, on the other hand, refers to the loss or detriment suffered in the patrimony due to the non-fulfilment of an obligation.

48 Hazard refers to a situation that produces a threat level whereby the occurrence of a potentially damaging incident is feasible.

Harsanyi, John C. 1956. Approaches to the bargaining problem before and after the theory of games: a critical discussion of Zeuthen's hicks', and Nash's theories. *Econometrica* 24 (2): 144–157.
Hobbes, Thomas. 2007. In *Leviathan*. ed. by Richard Tuck. Cambridge: Cambridge University Press.
Howard, John V. 1988. Cooperation in the prisoner's dilemma. *Theory and Decision* 24: 203–213.
Kalai, Ehud. 1977. Proportional solutions to bargaining situations: Interpersonal utility comparisons. *Econometrica* 45 (7): 1623–1630.
Knight, Frank H. 2002. *Risk, uncertainty, and profit*. Washington, DC: Beard Books.
Kreps, David, Paul Milgrom, John Roberts, and Robert Wilson. 1982. Rational cooperation in the finitely-repeated prisoner's dilemma. *Journal of Economic Theory* 27: 245–252.
Kraines, David, and Vivian Kraines. 1989. Pavlov and the prisoner's dilemma. *Theory and Decision* 26: 47–79.
Lewis, David K. 1979. Prisoner's dilemma is a newcomb problem. *Philosophy and Public Affairs* 8: 235–240.
Milinski, Manfred. 1987. TIT FOR TAT in sticklebacks in the evolution of cooperation. *Nature* 325: 433–435.
Morgenstern, Oscar. 1935. Vollkommene Voraussicht und wirtschaftliches Gleichgewicht. *Zeitschrift Für Nationalökonomie* 6 (3): 337–357.
Nash, John F. 1950. The bargaining problem. *Econometrica* 18: 155–162.
Nash, John F. 1953. Two-person cooperative games. *Econometrica,* 21, 128ff.
Nash, John F. 1996. *Essays on game theory* 1–8, 17–20. Cheltenham, Brookfield: Edward Elgar
Nash, John F., and Lloyd S. Shapley. 1953. A simple three-person poker game, annals of mathematics studies, 24. In *Contributions to the theory of games*, vol. I, 112–115. ed. by H. W. Kuhn, and A. W. Tucker. Princeton: Princeton University Press, 1950
Padilla Gálvez, Jesús, 2021. *El mentiroso. Genealogía de una paradoja sobre verdad y autorreferencia*. Valencia: Tirant humanidades.
Pareto, Vilfredo. 1906. *Manuale di Economia politica con una introduzione alla scienza sociale*. Molano: società Editrice Libraria. Trad.: *Manual of political economy. A critical and variorum edition*, vol. 2014 72–124. ed. by Vilfredo Pareto, A. Montesa, A. Zanni, L. Bruni, J. Chipman, and M. McLure. Oxford: Oxford University Press, 2014.
Pareto, Vilfredo. 1920. *Compendio di sociologia generale*. Florence: Barbèra.
Poundstone, William. 1992. *Prisoner's dilemma. John von Neumann, game theory, and the puzzle of the bomb* 8, 117. New York: Doubleday.
Von Neumann, John, and Oskar Morgenstern. 1944. *Theory of games and economic behaviour*, vol. 2004 15f, 85–90, 93f, 111f, 164, 177f, 188f, 205, 205, 216–219, 501, 512, 557, 616, 652, 714. Princeton: Princeton University Press.

Chapter 4
International Negotiation and Forms of Life

1 Introduction

When we analyse negotiations from the perspective of game theory, it is assumed that the participants act or proceed according to the model described in the context of games of luck. The assumption here is that participants would trust their counterparts and always act according to certain rules. In the present chapter, we will examine this view. While the negotiators' behaviour appears to be fixed in advance by the strategies of the game, it must also be stated that their actions are influenced by their forms of life and psychological peculiarities.

Before we start a negotiation, we need to gather as much information as possible about each negotiating partner. We need to specify the profile of the negotiator involved in each game. Every negotiator should be able to solve problems without resorting to confrontation or stalemate. As such, a suitable game theory should be able to offer a possible solution for such special situations. It must have regard to the specific characteristics of any negotiator and provide possible alternative courses of action. To achieve these ends, the negotiator should know the principles of the negotiation process, study the negotiators' opportunities and strengths, communicate effectively,[1] master a wide range of strategies and know how to develop interpersonal relationships.[2]

This chapter will define the concept of form of life and explore its application in the context of a negotiation.[3] We start with an analysis of our everyday use of the concept and compare its meaning to that of related concepts. The second part uses the categories of persistence to examine influence that forms of life can have on human interactions during negotiations.

[1] Negotiators should have a broad range and command of the language games used in a field of business and in interactions with business partners. This includes verbal as well as non-verbal communication.

[2] In interpersonal relations, negotiators are supposed to be empathetic and build trust.

[3] Gaffal (2022), 101–110.

M. Gaffal and J. Padilla Gálvez, *Dynamics of Rational Negotiation*,
https://doi.org/10.1007/978-3-031-49051-4_4

2 The Concept of Forms of Life

What do we mean when we talk about "forms of life"? In everyday usage, the notion of form of life refers to a whole series of heterogeneous phenomena in contemporary societies. It allows us to differentiate between urban and provincial forms of life, Spanish forms of life can be compared to English forms of life, unequal forms of life allow us to distinguish how bargaining is conducted in patriarchal societies and how it is conducted in companies where the employees' competencies are restricted. There are studies that deal with authoritarian forms of life and compare them to bourgeois forms of life, and so on. Scholars have long examined forms of life as important factors in understanding human social behaviour. The notion is also used to isolate particular forms of life to distinguish them from others whose characteristic features are not clearly analysed or cannot yet be described precisely.[4]

The concept of "form of life" became customary in Germany a century ago when Eduard Spranger distinguished between subjective and objective aspects of economic, aesthetic, theoretical and religious forms of life.[5] These characteristics referred to "ideal types of individuality" representing standardised ways in which individuals in a given society relate to each other and to the world in general.[6] Thus, forms of life describe groups of people living in a society in similar settings who share certain characteristics and behave in similar ways. At the same time, it proved possible to distinguish between very different types, whose commonalities are hardly noticeable at first glance. While in one case the relevant point of reference is a historical epoch, in the case of social groups it may be an organisational form of coexistence.

In our view, forms of life have a direct impact on negotiators' conduct, their modes of negotiation and patterns of behaviour. Therefore, forms of life are closely linked to actions in general, and to the language used by negotiators. While forms of life refer to historical processes in a society, sociological studies are restricted to the analysis of a specific behaviour over time. Thus, forms of life link human behaviour to a tradition. In the context of international negotiations, for instance, Chinese negotiators consider it impolite to get down to business without an initial phase of getting to know each other. They hesitate to do business with people they do not know, because for them, before demand can arise, value should first be generated at the personal level. It is therefore common for negotiators to enable personal relationships before starting to negotiate. As such, a Chinese negotiator's form of life differs greatly from that of a European businessperson.

Working in a company or institution implies sharing a particular form of life, which means making decisions according to specific guidelines and developing a way of dealing with matters that differ from the way other companies or institutions are organised. If we apply the concept of a form of life to socio-economic reality, then we do not only refer to economic structures, but also to how problems are posed

---

[4] Padilla Gálvez and Gaffal (2011, 2013a), *Introduction*; Padilla Gálvez and Gaffal (2013b), 9ff.

[5] Padilla Gálvez and Gaffal (2011), 7–16. Padilla Gálvez and Gaffal (2013b), 15f. 159.

[6] Padilla Gálvez and Gaffal (2013b), 111ff.

and developed, what procedures are used, who makes decisions, what elements play a role in decision-making, who conducts a negotiation, how the negotiation develops, what strategies are applied and who decides on them.

An employee who works for a company where paternalist, individualistic and authoritarian decision-making are entrenched will probably view a negotiation as a confrontation which ends in a winner-loser dyad.[7] Decisions will be taken individually and personally. In case of a mistake, responsibility will be delegated to other members of the company or to the opponent.[8] On the other hand, a negotiator working in an institution where the negotiation is conducted by a team, strategies will be discussed and shared solutions will be sought. For certain cultures, such as Chinese culture, it is relevant to maintain the same negotiation team during an entire negotiation, because, in the case of a change, negotiations will have to start from scratch.[9] The strategies belonging to the two types of business posited above not only bring into contact different characters, social structures, business structures, ways of working, etc., but are also carried out under the influences of forms of life rooted in distinct communities with their own histories.

Whatever the possible scope of forms of life, it reflects the orientations that determine individual patters of life and their shaping of a society.[10] In other words, the concept denotes how people live and what practices they engage in. As we have described two models of business coexisting today, we see a field of tension between static or traditional social models and business structures that are more in line with new forms of life. This discrepancy is not only a question of institutional changes in the social system, but also affects people's conduct in general. According to this perspective, the real impact of new forms of life in business, beyond the institutional changes involved, has been transforming the everyday life of citizens. Examples of this transformation are the integration of women into the labour market and the creation of institutions such as workplace nursery schools, where children from a very early age on are taken care of. This marks not only a transformation of gender roles and a new structuring of the family, but also a different way of approaching work and organising the daily plans.[11] New social structures have transformed family, social and institutional relationships. They have created new structures of cooperation and introduced conventions within social interaction including dress codes and a general new sensibility. These transformations have changed the forms of life differently across different societies. To the extent that forms of life are more open and variable, generating greater roots in social formations, they establish more stable structures and antecedents that constrain individual action. The forms of life that we have acquired shape our own options for action.[12]

---

[7] These companies tend to use zero-sum games: von Neumann and Morgensern, 1944 (2004), 85ff.

[8] Katz (2017), 22.

[9] Poza García, et al. (2010), 26–34.

[10] Padilla Gálvez and Gaffal (2011), 7ss; (2ª edición), 2013a, 7ss.

[11] Gaffal (2010).

[12] Gaffal (2016), 45ff.

3 Lifestyle Versus Form of Life

The vagueness of the concept of form of life, as attested by a first examination of its meaning and uses, suggests that it would be worthwhile to try to clarify the term by delimiting it from related concepts and phenomena. There are clear overlaps between the concept of form of life and that of life orientation in the sense of how one directs one's life towards an end or lives according to certain principles. These principles can be positively based on a certain selfless empathy, or, on the contrary, on a hedonism that seeks mere pleasure and gain at the expense of others. The concept of human behaviour refers mainly to individual phenomena; in contrast, when we refer to forms of life we are interested in collective phenomena in which people share certain practices in society. Thus, looking closely at the phenomenon described above, we would say that a form of life consists of someone getting rich at the expense of others because it is implicitly consented due to the existence of lax legislation. A pernicious law is the result of a form of life that allows or tolerates certain social groups to bear the harm or damage caused by a pattern of arbitrary behaviour previously established. Societies with such laws generally value positively the behaviour of the "clever" bargainer who enriches himself at the expense of a certain "useful fool" and place little value on the fact that unwary citizens are cheated through swindling. Such a form of life does not consider such cheating to be a swindle and therefore a crime defined by the expropriation of property or assets by deceit. Something similar is reflected in the board games on which game theory is based.[13] In fact, our personal behaviour is performed actively, whereas forms of life are based on passive attitudes that have been assimilated and have been performed for a long time, stemming from the past.

Citizens in certain societies are socialised into a form of life that has been prefigured previously. Social habits are related to forms of life in the sense that both share a propensity for regularity, stability and self-evidence. This becomes evident in the case of traffic jams, which occur in all societies at different times, and how they are solved. Studying these examples helps us to understand the relationship between habits and forms of life. Another difference can be noticed in the ways people greet each other in different countries. For instance, in German-speaking countries, rituals of greeting are rather formalised, and relations are at first distant. In contrast, in Spain people tend to greet the other as if he or she were a member of the family. When relations come to an end, in German-speaking countries there is a kind of established farewell procedure, whereas in Spain, there is no ritualised form, and one is often confronted with an abrupt end to the relationship. Certain practices are isolated to individual habits within the framework of a society, whereas the concept of a form of life refers to practices performed by a group. Forms of life, unlike habits, have a normative character.

Although we take it for granted that there is a close relationship between the expression "lifestyle" and "form of life", the difference could not be greater. A lifestyle refers to certain temporary practices and habits determined by a certain

[13] Nash (1996), 1ff.

contingency. This is why they are studied as phenomena linked to fashion. Lifestyles are subject to a certain mode of behaviour during a specific period of time in which similar individual or social interactions take place. The predominance of certain opinions based on cultural or epistemic canons shared by certain social groups can also be observed. Moreover, similar value attitudes are reproduced repeatedly. For instance, acting as an entrepreneur in a paternalistic style is considered outdated. In contrast, the role of a manager in a company structure in which Quality Management takes place, belongs to the form of life of a company and favours normative modes of behaviour among the employees of the company.

Customs are close to forms of life in the way they are organised and in the way they are binding. A custom is considered to have a strong normative bias, as it establishes implicit social rules that individuals observe voluntarily. Custom encompasses a variety of habits appropriate to a particular culture. To speak of customs, on the other hand, allows us to introduce a further feature that resembles forms of life, namely the ritualisation of individual actions passed on from generation to generation. Tradition is therefore related to forms of life in that it includes customs and habits. One could say that tradition is a long-lasting form of life.

There are certain similarities and discrepancies between forms of life and institutions that reveal important aspects of social action.[14] Forms of life, like institutions, are examples of social practices that have become habitual and have a normative character. But they differ in that institutional practices are firmly established and eventually become codified, whereas this is not the case with forms of life, where informal use prevails. Individuals living in a society share a form of life that is neither founded nor established, nor is it codified since it is mostly guided by implicit rules. Rather, forms of life represent the background and prerequisite for, or must be accommodated by, certain institutions. These institutions cannot be introduced into a community from outside without being linked to local forms of life, or where institutions, such as companies with an authoritarian boss, have become "outdated" and may even clash with existing legislation. Conversely, institutions become constitutive parts of forms of life and even facilitate or stabilise them. Institutions such as kindergartens or cafeterias, for example, make the modern nuclear family's form of life possible because the mother can work and does not need to spend so much time in the kitchen. Therefore, a form of life can also be understood as a set of practices that become institutionalised over time.

To sum up, there is a broad overlap between a form of life and culture, understood as the material conditions for the technological reproduction of human activity, which in turn organises social relations. The two differ from each other in that culture has traditionally been understood as a totality of social practices linked to a society and is therefore always asscociated with a certain uniformity and uniqueness. Forms of life refer to a plurality within the framework of complex societies. Forms of life are therefore intercultural, since they generate cultural expressions shared across different cultures.

---

[14] Gaffal (2016), 50f.

4 Central Elements of Forms of Life

The distinctions made in the sections above allow us to apply the notion of forms of life to different fields. The notion encompasses certain social practices that are carried out by specific social groups based on ritualised behaviour. It consists of habitual attitudes and modes of behaviour with a normative character that concern the collective action of society, even if they are not strictly codified or institutionally binding. This means that forms of life involve sets of practices that are interconnected and interrelated in one way or another. They may thus be considered collective institutions, i.e. informal norms based on socially shared practices. As habitually established formations, forms of life have both a passive and an active element. Every negotiator is born and lives in a form of life as in a structure that is given and established beforehand, even if one simultaneously creates it through one's own practice. In terms of the normative character of social cooperation pertaining to forms of life, the latter are based on regular practices whereby cooperation between individuals sharing the same form of life is essential.

To develop a more precise definition of how forms of life decisively influence negotiators and thus the negotiation process, in the following section we will address a range of complex issues that revolve around the social constitution of forms of life and assess the specific role they play. The relevant questions are the following: To what extent must a set of practices and beliefs be stable, meaningful, self-sufficient and exhaustive in order to be considered a form of life? Should we speak of what is implicitly preserved in a society as a form of life? Even the preliminary explanation of the terms suggests that a social formation can only be meaningfully described as a form of life if it can be identified as a stable and autonomous system, i.e. as an entity that is stable and self-sufficient in various respects and has some relevance for the shaping of human life. Temporary, superficial phenomena and sporadic practices are not forms of life, although in many cases it is difficult to establish a clear separation between the two phenomena. Therefore, the permanence of certain common features of a social formation becomes a criterion for determining whether it can be considered a form of life in which certain reiterative behaviours concur, for example, in a negotiation.[15] But, obviously, this proposal opens up new questions, for example, is understanding and demonstrating these criteria, in principle, only intuitively plausible?

The criterion of permanence is relatively easy to explain. Reiterative negotiation behaviour must show a certain persistence to be considered a form of life. Thus, the reiteration of certain registers in the performance of negotiators may exhibit some traits of a form of life. Thus, the destructive behaviour of certain negotiators towards elements presented during a negotiation by foreign negotiators may be based on a traditional view that undervalues members of that other culture. Such biased

---

[15] Repeated reference has been made to an intimate relationship between the proposal developed about behaviour as it comes to be studied by T. Parson and the proposal describing the behaviour of players in the game theory of J. von Neumann and O. Morgenstern. Cf.: Parson (1937). von Neumann and Morgenstern 1944 (2004), 638.

negotiators may therefore act unthinkingly, assuming that their negotiation partners are likely to provide a poorer quality product or service and thus systematically undervaluing the object to be negotiated. A negotiator's form of life influences his or her way of doing business. The permanence of certain characteristics is therefore an important factor in determining whether a behaviour is based on a form of life or a lifestyle or whether it is merely a temporary event. In practice, the question is how to arrive at an appropriate response, and in theory, how permanent such an attitude must be to be considered a form of life.

When we observe a certain repetition in the way a social group acts, then we assume that we are dealing with a form of life and we tend to attribute to it a deeper meaning than what would normally be associated with more superficial phenomena of individuals and community life. Should we speak of forms of life, therefore, only when certain central or important areas of negotiation are affected? But while there seem to be obvious examples of such practices and domains, it is not easy—and perhaps not even desirable—to distinguish them in material terms.

Clearly, when we refer to forms of life, we are not talking about isolated practices, but about interconnected and interrelated sets of practices. But to what extent must a set of practices be exhaustive for negotiation behaviours to be considered specific forms of life in each society? What degree of internal coherence can we expect when we carefully study the behaviour of another negotiator against our own predispositions and reiterative behaviours? It is very common for European negotiators, including Spanish ones, to discuss the object to be negotiated in terms of acceptance or rejection. Therefore, when they accept an offer, they generally behave politely, but if they do not, they tend to lose their composure. This marked duality is inappropriate when negotiating with other forms of life, as is usual in China. Few negotiators in Spain have learned to decline an offer politely or to break off a negotiation in a friendly manner. Quite the opposite is the case, often negotiations end by "burning all boats" in the manner of Hernán Cortés. This procedure may have its historical roots in Spanish history as a supremacist country that imposed the criteria in negotiations. However, in a world characterised by multilateralism and complex international relations, this conduct generates rejection.[16] Moreover, it is perceived as rude. This behaviour is widespread and can be observed repeatedly in state institutions and in the private sphere. It is much more efficient to end a negotiation with a goodwill agreement than to break ties with a negotiator who is likely to spread a negative view of the negotiator, which in turn can ruin a company's reputation.

If we compare the most relevant differences between lifestyles and forms of life, it allows us to evaluate the qualitative criteria of forms of life in contrast to more ephemeral phenomena. One paradigm of lifestyle is that it is shaped by trends, which allows us to delimit some important criteria when it comes to defining negotiation. The relevance of the social phenomenon of trend in the current context lies in the fact that it brings together criteria which, in crucial respects, are completely opposed to the concept of form of life. Three attributes can be extrapolated from the analysis

---

[16] See our comparative study of Germanic and Spanish cultures: Padilla Gálvez and Gaffal (2005).

of trends for current purposes: the dynamics of ternds are constitutively unstable, unrelated to reality and non-generalisable.

The first characteristic feature of a trend when applied to negotiation, which is what interests us in these pages, is its dynamic of change, the rapid disappearance and permanent interchangeability of the phenomena that constitute it. In the practice of negotiation, any strategy that is novel and suddenly spreads in the same way will not be characterised as trendy, if we believe in its permanence and objective justification. On the contrary, a strategy in negotiation is ephemeral and will disappear as quickly as it emerged, so we will call it a fad. A trend, one can conclude, is a phenomenon that is distinguished not only in fact but constitutively by constant change. Those who follow a trend in negotiation do not follow what has proved its worth but seek variety and thus constant novelty. The character of trends is such that the mere fact of novelty momentarily changes corresponding behaviour among negotiators. However, the fact that trends depend on a constant change of content, on a perpetual transformation, is not only a question of a contingent need for variation; the systematic reason is that trends always aspire simultaneously to conformity and differentiation.

Trend generates an attraction to conformity: once something has become trendy, there is a kind of imperative to follow it. Without this conformist trait, a transitory phenomenon of any kind is not a trend; it is only when others join in that something becomes a trend. At the same time—and this constitutes, to some extent, the paradox of trends—the strategy applied in the negotiation disappears as it has been introduced into the negotiation. Therefore, the negotiation in question is only really considered trendy as long as it can be considered a distinctive feature. In fact, trends can be interpreted in other forms of life as a bluff,[17] since such a strategy allows the real strategy of a negotiator to be hidden. Many cultures consider a trend to be nothing more than a ruse in the negotiation process.

With forms of life things are completely different. While it is difficult to assess quantitatively how long something must exist before it can be considered a form of life, there are good reasons to say that something that is constitutively unstable cannot be a form of life. Lifestyles share the dynamics of the changes typical of trends as described above.

5 Objections to the Proposed Model

The models developed so far in game theory have been detached from reality. On the one hand, the interaction proposed by von Neumann and Morgenstern is based on board games or invented models to develop a mathematical game theory. Others have developed models in line with the recurrent conflicts in their societies. All of them make the same mistake by developing game theory based on trends and without considering the negotiators' forms of life. In other words, the strategies designed in these games have followed unstable models, alien to reality because they are based

[17] Von Neumann and Morgenstern 1944 (2004), 186ff.

on the description of isolated actions that do not take the criterion of permanence into account. Our daily experience confronts us with social situations in our own societies in which we are forced to maintain a relationship with people belonging to other societies with whom we do not share common forms of life. Studies on game theory disregard centuries of domination, inequality and economic differences. All the models described follow a chess pattern in which two opponents of equal social rank bring their knowledge of the rules and their practical knowledge into play. The models used in chess assume a uniform status in terms of social conditions and status, where the players use similar strategies and techniques. In other words, it is formal intentions, mathematical interpretations or moral views that form the organising principle of game theory.

Game theory encompasses an unusual amalgam between a playful treatise, mathematical formalism and the illustration of the capitalist system.[18] Indeed, each example presents, in a display of ludic vocabulary, a wide and bizarre typology of characters that includes men, bluffing gamblers, race-baiters, prisoners, salesmen, advertisers, etc., who, curiously, all coincide in speaking the same language. Such a display of multiculturalism only as "culture" pertains to gambling and constitutes one of the most attractive aspects for future researchers interested in studying the sociology of game theory, which supposedly allows us to intrude into the eclectic spectrum of interests of the modern *homo ludens*.[19] However, all the supposed variety among "players" appears fictitious and imposed. The flaw of these studies is that many of them appear detached from reality.[20]

A recent case highlights how competitive forms of life can generate certain illicit behaviours. At the turn of the century, William R. Singer set up a network that falsified academic and sports results and bribed people linked to prestigious US university administrations to get the sons and daughters of influential businessmen and celebrities into these elite institutions. Singer believed that he could organise a "game" that would allow the wealthy to fraudulently gain admission to prestigious universities for their offspring. In doing so, he defrauded the state by falsifying grades, conspiring to defraud the tax authorities and laundering money through the organisation he had set up. It was interesting to see the justification he presented to the public prosecutor's office at the trial in 2023. In his view, he attributed this fraud to a form of life based on the principle of "succeeding by whatever means", in which bluffing and lying is tolerated if the desired result is achieved.

When game theory describes the tactic of "bluffing", one wonders how a person who has mastered the art of bluffing has managed to reach such a high position in society. How is it possible for a negotiating partner to deceive the respective other and thereby gain an unlawful advantage? After all, the negotiator has learned and practised this behaviour. This leads to the question: How can a bluffer use these tricks

[18] It is interesting to note the insistence on studying duopolies. von Neumann and Morgenstern 1944 (2004), 1, 13, 543. Nash (1996), 47ff.

[19] Huizinga (2014).

[20] See the work of A. Rapaport, in which the pros and cons of game theory are evaluated. Rapaport (1962), 108ff.

to sneak into the negotiating arena and thereby doom a company by putting many people out of work? One flaw in game theory is certainly that it does not reveal that all players who commit this kind of crime speak the language of the environment in which they were trained. The problem lies in the fact that many games described in this work take for granted that in asymmetric systems, the players, the parties, the negotiators do not use exactly the same language but think and express themselves with different words. Therefore, game theory is confronted with additional obstacles, which it has left unthematized.

These obstacles are expressed in the form of a dilemma. In fact, many of the subjects described in game theory presuppose the difficulty or impossibility of expressing themselves in the language used in the dominant game, which they have not learned to handle. Many moves in game theory assume aseptic players unable to communicate with each other and negotiate any issue as it comes to be described or presupposed in the model. What game theory is unable to thematise are the differences between players—or negotiators—who come from different social and economic backgrounds, who have been trained differently, who act and interact with each other according to certain learned rules or through lies. When game theory describes the modus operandi of the bluffing parties in the game, they are operating with a model that does not fit reality.[21] When they lie in a negotiation,[22] they usually change facts about the product or service, pay illegal sums to get a contract or block other negotiators. They also do not indicate, and therefore do not explain, how people from different cultures who speak different languages relate to each other. This is the problem we need to address next.

6 Unresolved Problems

Our use of the term "form of life" has been the subject of considerable controversy recently, although it is not a clearly defined concept. However, it has a long history, which has been characterised in different ways. On the one hand, it has referred to the manners and customs of societies in which certain social forms are linked to the way people communicate while on the other hand, the conventional aspects of socialisation anchored in language and non-verbal language have a direct impact on negotiation. From this point of view, forms of life are not understood as a moral vision of the world, but as a set of social techniques and practices that facilitate extensive social negotiation in a way that tends to minimise conflict.

From our point of view, the concept of form of life refers to a particular set of practices or behaviours that the parties display during a negotiation. These can be seen as patterns according to which the parties make their decisions. Forms of life are the basis for all actions in a negotiation. These patterns are adapted to the language games used by each negotiator or group of negotiators and their strategy or game to

---

[21] Von Neumann and Morgenstern 1944 (2004), 204–208. Nash (1996), 17–20.

[22] Padilla Gálvez (2021), 25ff.

achieve a given goal.[23] But how can trading partners belonging to different forms of life negotiate with each other, if their respective language games used for this purpose are different and prone to misinterpretation? Can negotiations contribute to mutual understanding, such that exchange through trade allows a rapprochement between parties? How can countries as different as Spain, Austria, China, Vietnam, Iraq, Kuwait, Ghana, Egypt, Argentina, Australia, Mexico, Canada, etc., negotiate with each other without essential misunderstandings arising?

Undoubtedly, the usual ways of behaving in China differ greatly from those considered appropriate in other countries. In fact, many deals do not materialise due to major differences, or they end after only a few meetings.[24] It is therefore advisable to find out as much as possible about the culture and business practices of one's business partner and acquire the appropriate intercultural skills. This is a dynamic process in which knowledge of another culture is applied to a given business situation aimed to reduce the unknown parameters and to facilitate negotiation. In addition to acquiring knowledge, this process also requires an openness to facing new situations and constellations that need to be constantly analysed and interpreted. There is widespread confusion among economists and sociologists about the assumption that all negotiations can take place between negotiators from different cultural backgrounds, since all negotiators share certain interests that can be negotiated within the framework of their forms of life. On the contrary, we believe that there are different forms of life, of a pluralistic nature, which constantly interfere with negotiation. People tend to seek information that confirms their own beliefs and to ignore information that contradicts their prior beliefs. To negotiate effectively, it is important to empathise with the other party's point of view, to be open to other points of view and to try to approach an issue from the other party's perspective. A more direct approach to understanding the other side is to explicitly discuss each other's perceptions. Each person should share their perceptions openly and honestly, without blaming or judging the other party.

In the process, it can easily occur that the parties act inconsistently with their own views. For example, one party may have preconceived perceptions and expectations about the opposing party. The other party may act in a way that directly contradicts these preconceptions, effectively conveying that the party is interested in an inclusive negotiation. Evidently, in his or her form of life, he or she has learned the game of keeping up appearances. This approach justifies a position based on previously learned principles and values in a negotiation. Thus, when we approach an issue from this point of view, it appears that the issue is less arbitrary and therefore may be more understandable from the perspective of the opposing party. In addition, this strategy allows negotiators to use communication strategies to build stronger relationships and thereby produce more efficient solutions.

Within the endeavour to understand the other, active listening occupies an important position. Certainly, listening is more than just hearing what the other party says.

---

[23] Padilla Gálvez and Gaffal (2013a), 8ff.

[24] For a detailed description of the leaderships style that is typical for Chinese companies, see Dolan and Kawamura (2015), 301.

Active listening involves paying close attention to what is said verbally and non-verbally. It involves periodically seeking further clarification of the other's behaviour. By asking the interlocutor what he or she means, one realises that negotiators do not simply follow a routine but take both sides' proposals seriously. However, too much information can be just as damaging as too little information. Before discussing an important point, it is necessary to clearly determine the issue in question.

The interrelationship between negotiation, language game and form of life allows a concise study of the relationship between two competitors with defined goals. In negotiation, the parties' objectives are unknown from the outset. The study of language games, the rules they apply in negotiation and the forms of life they presuppose help to clarify the negotiating strategies between parties. Following the rules, such as the controlled use of certain behaviours during negotiation, is determined by using language games and the knowledge of the negotiators' forms of life. It is therefore important that the rules that standardise the negotiation procedure and control the parties' behaviour are as transparent as possible. Forms of life encompass such elements as the history of a culture, the customs and habits of a given society that characterise the way of behaving. All of this is encapsulated in the language games that any speaker masters and with which negotiators must be familiar.

References

Dolan, Simon L., and Kristine M. Kawamura. 2015. *Cross cultural competence: A field guide for developing global leaders and managers*, 301. Bradford: Emerald Publishing Limited.

Gaffal, Margit. 2010. *Psychosocial and legal perspectives of marital breakdown*. Berlin, Heidelberg: Springer.

Gaffal, Margit. 2022. Negociación, teoría de los juegos, juegos de lenguaje y formas de vida. *Dókos* 29–30: 101–110.

Gaffal, Margit. 2016. Actions embedded in forms of life. In *Action, decision-making and forms of life*. Aporia, vol. 9, 45–59. Berlin, Boston: De Gruyter.

Poza García, David J., and Galán Ordax, José M. et.al. (2010). Emergencia de normas espaciales de negociación en el juego de demanda de Nash. In *4th International Conference on Industrial Engineering and Industrial Management XIV Congreso de Ingeniería de Organización*. Donostia-San Sebastián, 26–34.

Huizinga, Johan. 2014. *Homo ludens: A study of the play-element of culture*. Mansfield Centre, CT: Martino Publishing.

Katz, Lothar. (2017). *Negotiating international business. The Negotiator's reference guide to 50 countries around the world*. North Carleston, SC: CreateSpace.

Nash, John F. (1996). *Essays on game theory*. Cheltenham, Brookfield: Edward Elgar, 1ff., 47ff., 17–20.

Padilla Gálvez, Jesús., and Gaffal, Margit. (2005). *Spanienknigge. Sozioökonomische Einführung in die Interkulturalität*. München, Wien: Oldenbourg Wissenschaftsverlag.

Padilla Gálvez, Jesús., and Gaffal, Margit. (Eds.) (2011). *Forms of life and language games*. Frankfurt A. M., Paris, Lancaster, New Brunswick: Ontos Verlag, 7–16.

Padilla Gálvez, Jesús., Gaffal, Margit. (Eds.) (2013a). *Forms of life and language games*. 2nd edition. Berlin: De Gruyter, 8ff.

Padilla Gálvez, Jesús., and Gaffal, Margit. (Eds.), (2013b). *Formas de vida y juegos de lenguaje*. Madrid, México D.F.: Plaza y Valdés, 15f., 111ff., 159.

Padilla Gálvez, Jesús. (2021). *El mentiroso. Genealogía de una paradoja sobre verdad y autorreferencia.* Valencia: Tirant humanidades, 25ff.

Parson, Talcott. (1937). *The structure of social action.* New York, London: McGraw-Hill Book Company, Inc.

Rapaport, Anatol. 1962. The use and misuse of game theory. *Scientific American* 207 (6): 108–119.

Von Neumann, John., and Morgenstern, Oskar. 1944 (2004). *Theory of games and economic behaviour*. Princeton: Princeton University Press, 1, 13, 85ff., 186ff., 204–208, 543.

Chapter 5
Psychological Aspects of Negotiation

1 Introduction

When reflecting on the process of negotiating, we are not usually aware that we are faced with different situations in which we negotiate every day, whether in the family, with colleagues, when shopping or at the bank. All of these are situations in which we work with others to solve a problem together. For these everyday situations, everyone has acquired a repertoire of negotiation strategies. Negotiation is a process that takes place between people with different interests as they work towards a common goal.

To analyse negotiation, we need a conceptual scheme to describe the processes involved. In the previous chapters, we have proposed a general model with several key elements that apply to negotiation. We have placed the main emphasis on the study of negotiation strategies and have explored them from the perspective of game theory. However, in every negotiation there are psychological factors concerning the experience and behaviour of the participants that influence the process. Within these subjective and individual factors, recurring patterns of behaviour can be identified that typically occur in negotiations. We will examine the logic of these patterns and explain their functions and effects on negotiations.

Trade negotiations take place in various contexts and differ according to time and place or to an occasion, the object, its content, as well as to the parties involved and their interests and objectives. Negotiable items range from raw materials, products, tools or machines to services, ideas or innovations. The objectives of business negotiations may be to coordinate interests, negotiate contractual conditions or resolve conflicts. Apart from commercial dynamics, there are several psychological preconditions relevant to negotiators that are necessary for successful negotiations. In the following section, we investigate these conditions.

M. Gaffal and J. Padilla Gálvez, *Dynamics of Rational Negotiation*,
https://doi.org/10.1007/978-3-031-49051-4_5

2 Preconditions for Negotiations

All negotiation depends on multiple parties' degrees of willingness to coordinate their interests and objectives. In doing so, each party implicitly assumes that it can persuade the other(s) to accept an agreement compatible with his or her objectives. If they are convinced that they will achieve a better outcome by negotiating than by not negotiating or by choosing another option, then they are likely to apply the strategies described in the previous chapters. The parties' willingness becomes manifest through the interaction of their own interests and the other party's interest. From a game theory point of view, the willingness to negotiate can be described as a function between the values ascribed to self-interest and the other party's interests. Mutual dependence requires a minimum willingness to compromise in the case of conflicting positions. Pruitt and Carnevale view trust, power and overall positivity as essential elements of business relationships. They point to the mutual dynamics that trust usually brings about:

> Trust is the expectation that the other party will cooperate in the future. (…) But it is also an aspect of the relationship between parties, in two ways: it deals with how the parties perceive each other, and it is often (but not always) similar on both sides – if A trusts (distrusts) B, B is likely to trust (distrust) A.[1]

Reaching a compromise suggests that one abandons one's own demand for maximum benefit in favour of a common position.[2] The "Dual Concern Model" developed by Pruitt and Carnevale seems suitable to illustrate this mutual interdependence, as it combines the parties' interests to their negotiation strategies. It illustrates the negotiating situation from a one party's perspective in relation to the other party. The vertical axis shows the importance of one party's interests, while the horizontal axis represents the valuation of the other party's interests. Depending on the valuation of self-interests and the others' interests, the following five constellations arise (Fig. 1).

The authors explain the function of their model as follows:

> Rather than viewing self-concern (concern about one's interests) as a constant, as did earlier approaches, this model views it as a dimension running from weak to strong. When this concern is strong, one is willing to work hard for outcomes favorable to oneself; when it is weak, one is willing to let one's own interests slip. Other-concern (concern about the other party's interests) is also seen as a dimension that runs from weak to strong. Self-concern and other-concern are regarded as independent dimensions, rather than as opposite ends of the same dimension.[3]

We shall analyse each of the possible outcomes of the Dual Concern Model. The most favourable outcome is collaboration, where both parties give equal priority to their own and the other's interests. This cooperative attitude leads to maximum mutual benefit (a win–win game). In practice, it can also mean for business partners

---

[1] Pruitt and Carnevale (1993), 123.

[2] See the dynamics of game theory developed in previous chapters.

[3] Pruitt and Carnevale (1993), 104–105.

High -	Competence		Collaboration
Concerns about own		Commitment	
outcome - *Low*	Avoidance		Accommodation
	Low - Concerns about other's outcomes - *high*		

Fig. 1 Dual concern model

a reduction of the costs of joint projects, improvement of their market position or the stabilisation of their business relationship.

A compromise would be a combination of the parties considering each other's interests to be moderately important. This attitude results in a no-win, no-lose situation for both parties. The purpose of this strategy may be to maintain a business relationship despite a failed cooperation, or to exclude other equally strong business partners or to meet a pressure to win. By contrast, competition occurs when self-interest is prioritised and the interests of third parties are graded, one party will uncompromisingly pursue its objectives at the expense of the other.[4] This scenario typically arises in cases of a company's market dominance, in situations of emergency or when the stakes are high.

Accommodation appears as the most suitable strategy if a party gives more importance to the other's interests than to its own and is therefore willing to make more concessions than necessary. In that case, the party renounces intensive negotiations.[5] Although this means a loss in the present, a negotiator aims for a possible gain in the future. This strategy of gaining time is often used when the primary objective is to achieve a more far-reaching negotiating goal, avoid further losses, make the business relationship harmonious or force the trading partner to make concessions on subsequent projects.

Avoidance implies that if one of the parties considers both its own and the other party's interests as low, the motivation to negotiate is limited to the extent that a party withdraws from the negotiation table. As a result, neither trading partner gains anything.[6] This strategy comes into play when self-interest and the chances of winning are low, or when it is a matter of gaining time, taking a neutral stance or protecting a company's reputation.

A prerequisite for smooth transactions is that both parties behave in a fair manner. Any business relationship requires a minimum of trust among the parties. They

---

[4] This view has its correlate with the "win-lose" game in game theory.

[5] This is a typical case in which all players lose and therefore corresponds to the "lose-lose" game.

[6] Once again, a strategy set by the "lose-lose" game appears.

mutually assume that the other party will cooperate.[7] The reason why parties intend to build trust is to maintain a certain control over the negotiation. Moreover, the building of trust correlates with the choice of strategy and the expected duration of the business ties. The more profit a party aims to maximise from the start, the less willing they will be to cooperate and the shorter the business relationship is likely to last. If the goal is a long-term business relationship, it is important for the parties to detect what negotiation strategy the other party is using. However, even among cooperative negotiations teams the level of trust has an impact on their degree of cooperation. When trust is low, negotiators are reluctant to share information and first try to make sure that the other person is trustworthy.[8]

The interest in building trust varies across cultures. For countries known as *trust cultures*, including China, Japan, Korea and Thailand, trust is a central element. Thus, every negotiation is preceded by an initial phase in which the business partners get to know each other, and information is communicated implicitly. It is common that contact is only established on the recommendation of already known and trusted business partners. Trust is contextualised in a different way in so-called *factual cultures*, such as for instance the UK, Germany, Switzerland, the Netherlands and Scandinavia, where it is linked to competence and experience, meticulous scheduling and a high expectation of negotiation outcomes.[9]

Trust is built through communicative practices and has an affective and a cognitive component.[10] Both components appear together, but they do not always conform. In persuasive communication, the speaker seeks to act upon a matter by means of affective strategies. The goal of persuasion is not so much to evoke understanding or to exchange information as to interfere with the counterpart's attitudes and beliefs.[11] In the following section, we will examine persuasive communication as an interaction of affective strategies and mental attitudes.

3 Persuasive Communication

The art of persuasion is often used in commerce. Persuasive techniques aim to change the listener's attitude towards an issue.[12] Specially, trained salespeople use certain conversational techniques to weaken the customer's reservations by means of arguments and to overcome resistance by means of rhetorical tricks. Above all, they

[7] See Grice (1989), 248 on his theory of conversational implicatures and the difference between saying and opinion.

[8] Olekalns et al. (2008), 113.

[9] Powell (2012), 7ff.

[10] Paul Grice indicates that if a listener has confidence in the speaker who says it, then he himself should believe it: Grice (1969), 147f.; Grice (1989), 86f.

[11] Barwise and Perry (1983).

[12] Batinic and Appel (2008).

must convince customers emotionally.[13] This can be achieved through argument, by presenting a credible authority figure or by basing arguments on analogies. Experiments have shown that potential customers tend to pay little attention to arguments, while they are more receptive to affective stimuli. This can be done through humour[14] or false information,[15] such as that presented in talk shows or fake documentaries.

The boundaries between influence, manipulation and deception are fluid. Influence becomes manipulation when information has been altered by additions and omissions in such a way as to mislead the receiver about a fact. The transitions from persuasive sales talk to deliberate deception is smooth. However, for the deception to succeed, the sender needs to know the recipient's weak points. To do this, psychological tricks are used that are intended to disrupt the recipient's rational thinking. During brief interactions, the liar takes advantage of the recipient's credulity or politeness and continually oversteps the other person's limits or scruples.[16] One tactic is that of consistent action and reciprocity, in which the liar makes a small gift to his counterpart and expects a favour in return. The interlocutor is taken by surprise and, guided by courtesy, returns the favour. These tricks are intended to deceive the recipient and lead him to reactions that he has neither planned nor thought over.

A loss of trust and credibility always has a detrimental effect on a business relationship and may be associated with increased costs. If a supplier lies to his business partners about the quality of the products, then it is only a matter of time before the buyer recognises the false information and insists on compensation. Consequently, overall transaction costs increase because lawyers, accountants or mediators have to be called in to settle the resulting damages. The following section is devoted to the question of when deception in business is legally punishable.

4 Legal Consequences of Deception

In practice, the tendency to deceive others is as old as our cultural history. Hermes, the messenger of the gods, was not only the patron saint of travellers, merchants and shepherds, but also the protector of thieves and a symbol of oratory and magic. In professional life, deception is commonplace, in some sectors it is even part of the

[13] During negotiations, the decision whether or not to reach an agreement depends in part on emotional factors. Negative emotions can provoke intense and even irrational behaviour and can cause conflicts to escalate and negotiations to break down, but they can be decisive in securing concessions. On the other hand, positive emotions often make it easier to reach agreement and help maximize joint gains, but they can also be decisive in achieving concessions. Discrete positive and negative emotions can be strategically displayed to influence task and relationship outcomes and may develop differently for different forms of life.

[14] Grice (1989), 92.

[15] We have previously analysed the factor that determines bluffing and lying. Note that von Neumann and Morgenstern consider it paradoxically as another ingredient of "rational" negotiation.

[16] The liar constantly abuses occasional expressions in his speech in order to introduce his lies. Padilla Gálvez (2021), 340.

sales strategy. Proving deception, lies or bluffs in negotiation has its difficulties. For instance, a seller who misinforms a buyer about his products can be prosecuted once the contract is signed. However, the situation is different for a buyer who, during sales negotiations, claims that there are cheaper offers available, which do not exist. If the seller reduces the price of his goods and consequently sees his profit margin reduced, the buyer cannot be legally pursued. On the other hand, deception can have legal consequences that are often not entirely clear to the deceiver. In the next section, we will see how deception can be legally punished.

We have indicated that a loss of trust always has a negative impact on the business relationship and is often associated with increased costs. Business negotiations follow an established protocol and have clearly defined objectives. When trading partners enter a business relationship, they agree to exchange goods or services for money. For a contract to be valid, several conditions must be met. The seller must make an offer that the buyer accepts. Both must be willing to be legally bound and any performance must be balanced or compensated by consideration. It must be clear who is the principal or who acts as agent in his stead. The parties to the contract must be of legal age and capacity, and the terms of the contract must be precise and clear to avoid misunderstandings.

From a legal standpoint, misleading a business partner is a crime. In the *Uniform Commercial Code* (UCC), this is specified as fraud and is regulated in several sections, in which different forms of fraudulent conduct are described.[17] For example, an employer is liable for fraud if one of his employees unlawfully transfers property to a third party, or if payment instruments are fraudulently transferred to fictitious payees.[18] If a person is deprived of control over his or her tangible (or electronic) transferable document by way of misrepresentation, fraud, accident, mistake, duress, loss, theft or conversion, this is legally punishable.[19] In a related section, we find deception defined as someone's intention to mislead another person's conception regarding the facts of a matter or object. Deceit with the aim of illicit enrichment is regulated in the Criminal Code. Whoever, with the intention of obtaining an illicit patrimonial benefit for himself or for a third party, damages the patrimony of another person by creating or maintaining an error under false pretences or by falsifying or suppressing true facts will be punished with a custodial or pecuniary penalty.[20]

Deception in the form of simulation of false facts can become manifest in various ways, either by conduct or by failure to comply with the obligation to inform, both of which are considered breaches of duty. However, to punish fraud, it must be possible to prove that the deception was related to specific facts or conditions. For this, three types of deception are distinguished: first, what is considered express deception occurs when the deceiver makes either verbal statements, puts a statement in writing or gives to understand something by gestures. This is sometimes done in the form

---

[17] Uniform Commercial Code (1952a).

[18] Uniform Commercial Code (1952c). Employer's Responsibility for Fraudulent Indorsement by Employee.

[19] Universal Commercial Code (1952e). Impostors, Fictitious Payees.

[20] Universal Commercial Code (1952b). Rights Acquired by Due Negotiation.

of falsified documents, or, for instance, by manipulating a vehicle's odometer or by a seller's gesture of assent when asked whether the product is really the specified branded product. Second, implied deception exists if the deceiver's general behaviour is to be understood as a tacit statement of fact. Third, there is deceit by omission, where a partner to a transaction was obliged to avoid, eliminate or rectify an error but did not do so. This refutes the false assumption that inactivity produces no legal consequences. On the contrary, a deceiver's omission of an obligatory action legally binds him in a negative way.

If a deceiving business partner intends to avoid legal sanctions, it is tactically more profitable for him to carry out the deception in such a way that the other partner does not directly recognise it as deception. As a result, the other party may be induced to make decisions that benefit the deceiver. The difficulty is that the inducement shifts the action to the side of the aggrieved party and diverts attention away from the deceiver. Since the deceiver's intent occurred implicitly and indirectly, the burden of proof falls on the deceived party. Another typical feature of deception is that it is often not carried out by a single person, but coordinated by a group of deceivers. Sometimes it is even anchored in a system, for instance in a company or an institution. In such cases, it is often impossible for the affected person to recognise the deception initially, as there are no suspicious signs of deceit. Thus, the deception only becomes apparent after the transaction has taken place or at a time when it can no longer be reversed. To sum up, all tricks in negotiation have in common that they are intended to weaken the counterpart's capacity of decision-making. They are aimed at inducing cognitive distortions and tend to lead to erroneous conclusions. In the following section, we will list the most common tricks that occur indirectly in negotiations and which are therefore difficult to sanction.

5 Negotiation Tricks

Several similar terms are used in connection with deception. The expression "trick" points to a practice meant to deceive or defraud. The term "cheat" is associated with being untrustworthy, misleading and dishonest, and it describes the use of a dishonest or unfair action to gain an advantage. For example, an "accounting trick" is the manipulation of a company's financial statements to make them look more positive than the actual performance would suggest. Compound "dishonest dealings" refer to transactions that are carried out under the table and without being officially documented.

Generally speaking, tricks used in negotiations have two objectives: to disguise a party's intention and to mislead the other party about the valuation that the first party ascribed to an object of negotiation. In this procedure, the negotiator's ingenious ability to make a profit plays an important role. We must not lose sight of the fact that some tricks fall within the scope of the crime and therefore are regarded as fraudulent behaviour, are contrary to the law and carry a penalty. The following tricks are frequently used.

Lying

A lie is a false statement that is deliberately made to deceive someone. In negotiation, its object is usually the obtaining of unfair advantage.[21] For lies to be effective, they must be plausibly expressed and give the appearance of accuracy and truth.[22] The interlocutor can probably reveal this direct form of lying by addressing the negotiating partner directly about the matter. The best way to avoid lying is to find out about the other party's strategy, goals, obstacles, failures, etc., beforehand. To find out if the other party is lying, it is advisable to ask direct questions and, if the answer is still unclear, to inquire further. The more the other party is pressed by questions, the more verbal or non-verbal signs of inconsistency will come to light.

The situation becomes more complicated when the deceiving party deliberately withholds relevant information, which has adverse consequences for the deceived party. Silence or omission plays an important role in this respect because the intention of the deceiving party is not openly manifest and therefore cannot be proved.[23] Often, this type of omission can only be determined retrospectively and when important transactions have already taken place, by examining the state of knowledge of the deceiving party at the time in question.[24]

False Non-Verbal Messages

A negotiator needs to know the basic non-verbal elements of how other negotiators are communicating.[25] This knowledge can help significantly in the communication process.[26] By being aware of the inconsistencies between verbal and non-verbal communication between negotiators, reconciling them can help negotiators improve their final decisions in a negotiation. The intention of sending false non-verbal messages is to mislead a trading partner about one's intentions. Thus, a negotiator may show impatience through certain facial expressions, gestures and body postures, thereby pressuring his interlocutor to accept a disadvantageous offer without further consideration. An attempt is made to catch the business partner, who is willing to cooperate, by surprise. In the so-called flinching, one of the parties exhibits a strong,

[21] Padilla Gálvez (2021).

[22] Grice (1989), 55 ss.

[23] Levinson (1997), 299 s.; 326 ss.

[24] It should be noted that omission is considered an offence which produces a result caused by the negotiator's failure to prevent, thereby violating a legal duty.

[25] Levinson has included among non-verbal signals the gaze, i.e. eye contact between negotiators (Levinson 1997, 72, 95, 297 and 302), pauses (Levinson 1997, 299f. and 326ff.) and prosody, linked to intonation, lengthening and stress (Levinson 1997, 36).

[26] Barwise and Perry (1983), 277–279; 281–286 have developed a model of verbal communication to distinguish it from non-verbal communication.

negative physical reaction to the other party's proposal in the form of a facial expression that signals discomfort. A similar trick is to confuse the interlocutor by giving inconsistent and contradictory signals, to induce the other party to take rash actions.

Caution is advised when negotiating with interlocutors from different countries, as the meaning of non-verbal signals may vary according to culture and forms of life. For example, Asian business partners have subtle body language that remains ambiguous to people from Western cultures. On the other hand, certain gestures can mean exactly opposite things in different countries, as will be explained below.

Incorrect Valuation

This strategy consists of concealing the value of the object of negotiation from the negotiating partner to take advantage. On the one hand, a party may misrepresent the value of a negotiated object to the company to obtain it more cheaply. On the other hand, one of the parties may also lead the other party to believe that the value of a product is higher than its actual value to sell it at a higher price or justify a higher price. To avoid this, the value of the product should be accurately determined at the bidding and pre-information stages. In international negotiations, caution is advised, because the principle of good faith is contextualized in different ways in the legislation of each country.

Pretending Disinterest or Lack of Knowledge

This trick consists of one party responding to an opening offer from the other party with disinterest to get a better offer. In the previous chapters, we have referred to the causes and motives why a negotiating partner shows disinterest. In earlier chapters, we pointed out the importance of feigning disinterest in negotiations. Now we emphasise that instead of making hasty concessions, it is advisable to ask them under what conditions they would be interested in a deal. It is recommended to present a new, lower opening offer to find out if there is really no interest.[27] This tactic can be particularly effective in the negotiation phase to get the other party to make better concessions. If the other party has already included concessions in the list of agreed contract terms without having provided anything in return, then it is advisable to withdraw the concessions and only make a new offer once the other party's terms have been clarified. However, this strategy is less common in countries with rather short negotiation processes, such as in Australia, Austria, Germany, Switzerland and Nordic countries.

---

[27] Katz (2017), 73.

The technique of feigning ignorance about an issue consists of pretending to be a layman,[28] someone who does not know the details of the negotiation and wants to get the other party to take charge. This trick is associated with the expectation that the party will assume some sort of protective role and make more concessions than in arm's length negotiations. This includes the typical behaviour of not taking a clear position throughout the process, not sticking to the negotiation steps, and repeatedly changing one's point of view during the negotiation. A similar trick is to confront the other party with open-ended questions that are left off the agenda to elicit additional information. The intent is to obtain more information about the objectives of the negotiation and the other party's position on them. If it is suspected that a party is using this strategy and is dissembling in this way, the party should be asked to clearly communicate its interests and to respect the timetable and structure of the negotiation process.

Causing Confusion and Distraction

This trick consists of a party's aim to overwhelm the other with too much information so that it becomes impossible or too difficult to determine which information is relevant.[29] As a result, the party loses track of the amount of irrelevant or superfluous information received, which also has an impact on the objectives of the negotiation, which become unclear. Paul Grice has developed conversational maxims in which relevance is included. Following the maxim of relevance maintains the thread of conversations and helps to avoid random conversations that lack continuity.[30] This maxim also helps us to understand expressions in conversation that may not be obvious at first. If the negotiation follows this maxim, then negotiators should only say things that are relevant to the negotiation.

By creating a smoke screen, a party may intend to camouflage a deficiency in a proposal when presented to a business partner. Once a party has discovered the weak point in the other party's argument, the latter uses distraction to divert attention from it. To do this, the party makes unexpected demands or points to a sudden misunderstanding of the agreed terms. A similar trick is to make several small demands

[28] Pretending is an activity that occurs in a variety of circumstances, but all are aimed at inducing someone to perform a particular action. There is no consensus when it comes to clarifying how feigning works, since scholars fundamentally disagree on what "feigning" is. Consequently, they also disagree on the mental states that enable pretence. Representational theories argue that pretending requires the innate mental state concept of pretending. As such, pretending would be something like acting out one's representations. Behavioural theories, on the other hand, hold that participating in the simulation game requires an "as-if" behavioural process. The problem is not how to understand "pretending" but extends to "how to recognise that one is pretending".

[29] Grice (1989), 86–88.

[30] Pragmatic studies are based on the universal search for relevance. P. Grice introduced the maxim of relevance (Grice 1975, 64–75), systematized by D. Wilson and D. Sperber, indicating that relevance in a given context has to be approached in terms of contextual implicature (Wilson and Sperber 1986, 243–258).

towards the end of the negotiations to obtain further unconsidered concessions from the other party.

Threat and Intimidation

In all legal codes, intimidation and threat are considered an obstacle to concluding a contract. However, threat is sometimes applied indirectly, so that it is not directly recognised as an illicit tactic. For instance, a kind of weak threat occurs if one party puts pressure on the other by arbitrarily setting deadlines. The skilled negotiator should react to this tactic by making it clear in advance how they will react to the expiration date.

The more credible a particular threat is the more effective it tends to be, whereby credibility depends on a party's past experience with threat and on the question of who exercises it. Although pressure techniques may be effective in cases where the other party refuses to negotiate and it is the only means to bring the party to the negotiating table, they have several disadvantages. Pruitt and Carnevale underline the negative effects of threats in the following quotation:

> However, threats have their downside because they tend to generate resentment and resistance. People are not happy when they are forced to take action. It diminishes the freedom of choice and violates their sense of justice (...). As a result, they often try to escape or to build some form of counterpower against the threatener, another form of contentious behavior.[31]

On the other hand, if a threat has no consequences, it loses its significance as a negotiating tool. By demonstrating its own size and possibilities, one party tries to intimidate the other to achieve a better negotiation result. The more powerful party will flaunt its image or invite the other party to impressive locations to intimidate it.

Bluffing

In previous chapters, we have examined the function of bluffing in the framework of game theory. This trick occurs when a party presents an irrelevant point as if it were extremely important and demands a high consideration in return. The better informed the other party is, the more likely it is to recognise this trick and counter it. The line between a bluff and a lie is a fine one.

---

[31] Pruitt and Carnevale (1993), 31.

False Concessions

One party makes steep demands, arguing that it will offer concessions in return towards the end of the negotiations. However, these proposed concessions are in fact not what they seem to be, since they are either included in the offer, have already been agreed upon or do not require the party to incur any additional costs. The trick is aimed to induce the other party to make a greater concession than it has originally planned. In countries that reject intense bargaining, this trick can lead to success, since the aim is to limit bargaining by compromise. The best reaction to exaggerated demands is to keep calm, explain why their proposal is not feasible and follow the agenda.

The Good Cop, Bad Cop *Strategy*

The trick is for one or more people in a negotiating team to play the role of *bad cop*, forcing the other party to make concessions by making high demands and adopting a hostile attitude. To keep the negotiations moving forward, the rest of the team acts rationally and cooperatively, playing the role of the *good cop*. Once the other party has made concessions, the bad cop makes further demands until the team has achieved its negotiation goal.[32] This trick is particularly successful in countries where the business relationship is of only moderate importance and the focus is on the outcome of the negotiation. Such countries include the United States, Germany, Belgium and the Netherlands. In Asian countries, however, this tactic can cause confusion and lead the other party to break off the business relationship, as they conclude that an agreement will be impossible, and they do not wish to create a conflict.

Straw Man

In this trick, the decision-maker stays out of the negotiation by sending others with limited authority. This allows the negotiator with limited authority to respond to the other party's demand by saying that he must ask permission to move forward in the negotiation. This creates a dynamic like that of the good cop, bad cop strategy described above. In one party, this creates the contradictory impression that the strategy in the negotiation room is cooperative, but nevertheless depends on an absent third party who appears uncooperative. They conclude that the decision is made autocratically by a person who is not present at the negotiation. The aim of this tactic is to cause the two parties in the negotiation room to cooperate in satisfying the demands of the absent decision-maker.

---

[32] For a detailed description of this trick see Katz (2017), 75ff.

Incoherent Behaviour

In this trick, one of the parties displays irrational behaviour by switching between cooperative and hostile comportment, sending false non-verbal signals or presenting irrational arguments. The same occurs if a party appears enthusiastic about a proposal and suddenly changes to a position in which it does not show any reaction. The aim of this tactic is to induce the other party to make unilateral concessions to stabilise the situation. An unexperienced negotiator could be tempted to overreact by facilitating additional information or by offering concessions to get the negotiation back on track. If a party is confronted with this trick, it should remain calm and not respond to it, but rather continue the negotiation as usual following the negotiating guidelines.

In negotiations between parties from different cultures, misunderstandings may arise since what is considered rational or irrational varies according to their forms of life and culture as well as to the objective of the negotiation.[33]

6 Deception and Body Language

Negotiations can be compared to a game of chess. Each player makes the move depending on how he or she assesses the overall situation. Each party assesses the state of the negotiation process from its own perspective and neither party asks whether their evaluation coincides with that of the other party. Cognitive and affective elements are involved in every negotiation. Interactions in negotiation involve many complex issues and contexts that interlocutors process in different ways. Why do conflicting signals occur? During the negotiation process, interactions of various kinds take place in which the partners have to reconcile the information they receive with their respective negotiation strategies. Processing this information often generates contradictions that must be resolved effectively to continue negotiating. These contradictions can take the form of a mismatch between the verbal messages and the non-verbal signals. Rationally, these situations are easy to control, but emotionally, reactions are not always controlled. If a negotiation lasts several hours, the control of negotiators' facial expressions, gestures and posture tends to decrease.[34] If someone is in tension for a long time, it is sometimes expressed in involuntary gestures. These physical signs point to a person's emotional state. Verbal messages are not always in line with a person's thoughts and feelings. For example, a party, for whatever reason, could make an initial offer that they already know is unfeasible. To find out

[33] The terms "rational" and "irrational" do not have universal meaning but their meaning depends on how rationality is contextualised within a form of life.

[34] We have already indicated above that the book published by St. C. Levinson and entitled "Pragmatics" includes in the framework of non-verbal signals the gaze, i.e. the eye contact between negotiators (Levinson 1997, 302), pauses (Levinson 1997, 299f. and 326ff.) and prosody, linked to the intonation, lengthening and stress of sentences or words (Levinson 1997, 36).

what a person really thinks, we want to focus on cases where the linguistic and non-linguistic aspects of communication appear incongruous or where a contradiction arises. In a negotiation situation, this may indicate an intention to deceive on the part of the interlocutor.

Paul Ekman and Wallace Friesen offer a detailed description of how non-verbal cues are related to deception.[35] They examined the effort that a deceiving person must make to avoid leaking information or revealing bodily cues to the intended deception.[36] Generally, negotiation is a situation of mutual distrust in which both negotiators can symmetrically assume both the role of the deceiver and that of the detector. This dual role causes an inner tension, whose suspicious giveaways in body language need to be controlled. Among the non-verbal signs, facial expressions are the most revealing; followed by posture, including arm and hand movements, while leg and foot movements tend to receive less attention. To avoid revealing signs, the deceiver must pay special attention to control facial expressions and the movements of the body and upper extremities. Among the reasons why control of non-verbal signals can fail may be that the person is not aware of the effect of certain movements because he has not received feedback on them so far. Another reason is that a certain non-verbal sign could be considered normal in one person's form of life and therefore need not be avoided.

The following section provides an overview of most common non-verbal signals and their meanings. Since a gesture can have different meanings depending on culture and country, it will be explicitly stated if there is a cross-cultural difference. In the following characterizations, we will first introduce the non-verbal sign, then its universal meaning and finally we point out the different interpretations depending on the cultural contexts.

Facial Expressions

Some examples of incongruence in body language are the following: a laugh that does not match the situation may be a sign of nervousness or discomfort. When this occurs, it may be good to probe by asking questions to discover a negotiator's true feelings. If a person smiles and at the same time raises his or her hands in a closed position, it may indicate a negative attitude or frustration. If one detects contradictory signals of body language in the negotiating partner, it is helpful for negotiation partners to spend extended time together in a comfortable setting outside the negotiation room. Knowing how each partner communicates non-verbally outside the negotiation environment helps them detect incongruities between verbal and non-verbal communication. Here are some examples of facial expressions and their meanings.

---

[35] Ekman and Wallace (1969), 88ff.

[36] Ekman and Wallace (1969), 94.

Eye contact is a meaningful gesture. However, the type of visual attention varies from culture to culture. Depending on the intensity and duration, the act of looking or being looked at is associated with dominance, hierarchy, but also with attention and affection. In all countries, looking at the interlocutor is a sign of respect, whereas gazing is often seen as a sign of dominance or aggression. Looking up with the eyes implies dissatisfaction. Tight lips and tension around the mouth indicate rejection or resistance.[37] If a person uses large gestures to eloquently defend his or her position, there is a suspicion that the position is dishonest or not meant seriously. In the United States, children are told to look adults in the eyes when talking and therefore for Americans it is a sign of respect. In Arab countries, gazing is quite common and inspires sincerity, interest, and courtesy. However, one should be careful in allowing for prolonged eye contact in an intercultural negotiation because it is considered rude in Asian countries.[38] If a person looks down, it creates the impression of uncertainty or embarrassment. However, the same facial expression transmits respect in Turkey.

Gestures

Sometimes a negotiator describes an issue in positive words, but his body language reveals negative signs. Whereas resting the chin on one's hand indicates scepticism, doubt or uncertainty, resting the chin on both hands expresses that a person is bored. If someone inclines his or her hand slightly while listening, this shows that the person is interested or patient. Shaking one's head horizontally generally means negation; however, in India, Sri Lanka or Bulgaria, it expresses consent. Nodding one's head is known to stand for agreement, whereas in India and Sri Lanka it stands for denial.[39] Pointing at others has an aggressive connotation, but in Asian countries it is considered an outright insult. If a person crosses his or her arms this transmits distance or contempt. Playing with a pen reveals impatience or doubt. Movements of the hands are particularly informative; for instance, hand-wringing expresses rejection and resistance. If someone holds on to the armrests of a chair, to a writing table or a reading desk, this reveals tension and uncertainty. If the hands are put in prayer position, this means that this person is asking for something. Rhythmic finger tapping on the table or looking at the watch intuitively tells us that the person is impatient. If the fingertips are in a hedgehog shape it reveals resistance, whereas putting the fingers together like a rhombus transmits sovereignty and patience. On the contrary, clenched hands are a sign of aggressiveness and resistance. If someone rubs the chin, this means satisfaction.

There are some symbolic gestures that do not require verbal information to convey a message. Although these gestures are internationally known, they can have a specific meaning in individual cultures. For instance, a clenched fist with thumb

---

[37] Matsumoto et al. (2013), 34, 42.

[38] On the interpretation of eye contact in different countries see Zhou and Zhang (2008), 92.

[39] Gesteland and Gesteland (2010), 72.

extended upward denotes a positive response in many countries and its use as a positive assessment of a content has increased even more due to the spread in social networks.[40] In contrast, it is considered an insult in Iran. Great caution is advised with the so-called finger circle in which the thumb and forefinger form a O-sign for "Ok", which means "approved" or that everything is going well. On the contrary, in France and Belgium it has the opposite meaning, as it indicates that something is considered worthless. In Latin America, Russia and Eastern Europe the same gesture has a sexual reference, and in the Middle East it denotes homosexuality. In Japan it means that people want to talk about money. The movement by which the earlobe is pulled means uncertainty, whereas in Portugal it is a response to delicious food and in Italy it is a sexual innuendo. If someone pulls the nose, which is also known as the "Pinocchio effect", it points to deception or lying. On the other hand, slightly tapping one's nose with the index finger points to confidential information in England, whereas it means caution in Italy.[41]

The gesture of repeatedly tapping one's head with the index finger should be used with caution. It has opposite meanings depending on the country or culture in which it is used. Whereas in France, Italy and Germany, the gesture is used to indicate that someone is not to be taken seriously. In Germany, it can even be legally punished by the person to whom it is directed and result in a fine. In contrast, in Spain it is used to indicate that a person has either acted wisely or is intelligent.[42]

Power Poses

The way in which negotiation partners position their bodies in relation to each other can show their receptiveness towards the other person's messages. It is important to note whether receptive negotiators maintain eye contact or smile. This may be a sign of the negotiator being more interested in the person than in what is being said. Conversely, non-receptive negotiators tend to maintain little or no eye contact. Their eyes may be narrowed, jaw muscles clenched and head slightly turned away from the interlocutor. To show receptivity, negotiators should extend their arms or relax them on the table. Negotiators convey little receptivity when their hands are clenched, crossed, placed in front of their mouths or rubbing the back of their necks. Here are some more postures and the meanings associated with them:

A rigid posture tends to stand for inflexibility and immobility.[43] Bouncing legs intuitively express uncertainty, nervousness and lack of poise. If someone sits in a slightly laid-back position this transmits an attitude of laziness or boredom. Conversely, someone's leaning forward on the chair may point to a certain degree of aggressiveness. If someone uses the so-called mirror technique by which he adopts

[40] Matsumoto et al. (2013), 78.

[41] About the meaning of hand gestures in different cultures see DuFrene (2014), 71.

[42] Gesteland (2012), 87ff.

[43] McKay et al. (2018), 61f.

a similar posture to his interlocutor, this inspires caution. If a negotiator keeps the hands at stomach level, it expresses safety. If a person crosses his legs, there is always the risk that the sole of his shoe will point to the other person, which is generally considered disrespectful, especially in Asia, the Middle East and in Islamic countries.

Responsive negotiators tend to sit with their legs together or with one leg slightly in front of the other. When standing, they distribute their weight evenly with their body leaning slightly towards the interlocutor. They tend to appear generally relaxed, with their hands open and palms visibly exposed. In contrast, less or non-responsive negotiators tend to stand with their legs crossed, pointing away from the interlocutor. They may sit on the edge of their chairs, unbuttoning their suit jackets and with their bodies leaning towards the speaker. Some power poses are used by negotiators to convey the impression of importance and dominance. Here are some significant poses and their meanings: wide and sweeping hand movements are used to convey self-confidence. Standing with legs open and hands on hips may give the impression of dominance.[44] A person's being well seated with arms open and legs slightly open expresses a positive attitude towards the other person, whereas a person's arms resting right and left on the backrest inspires dominance and overall control.[45] If someone leans on a table with one's hands, it means that one is claiming power.

Being able to understand the non-verbal signs that negotiators send can help negotiators significantly in the communication process. If we become aware of the contradictions that may exist between a person's verbal and non-verbal communication, we can address inconsistencies, clear up any ambiguity and avoid possible misunderstandings.

References

Barwise, Jon., & Perry, John. (1983). *Situation and attitude.* Cambridge Mass: The MIT Press, 277–279, 281–286.

Batinic, Bernhard, and Markus Appel. 2008. *Medienpsychologie.* Berlin, Heidelberg: Springer.

DuFrene, Debbie D., ed. 2014. *Communication beyond boundaries.* New York: Business Expert Press.

Ekman, Paul., & Friesen, Wallace V. (1969). Nonverbal leakage and clues to deception, Psychiatry. *Journal for the Study of Interpersonal Processes* 32 (1): 88–94.

Gesteland, Richard R., & Gesteland, Mary. (2010). *India – Cross-cultural business behavior. For business people, expatriates and scholars.* Denmark: Copenhagen Business School Press.

Gesteland, Richard R. (2012). *Cross-cultural business behavior. A guide for global management.* Frederiksberg: Copenhagen Business School Press, 87ff., 172.

Grice, Paul H. 1969. Utterer's meaning and intentions. *The Philosophical Review* 78: 147f.

Grice, Paul H. (1975). Logic and conversation. In *The logic of grammar*, eds. D. Davidson and G. Harman. Encino, CA: Dickenson, 64–75.

Grice, Paul H. (1989). *Studies in the way of words.* Cambridge Mass.: Harvard University Press, 92, 248, 55ff., 86ff.

[44] Gesteland (2012), 172.

[45] Matsumoto et al. (2013), 83.

Katz, Lothar. (2017). *Negotiating international business. The negotiator's reference guide to 50 countries around the world*. North Carleston, SC: CreateSpace, 73.

Levinson, Stephen C. (1997). *Pragmatics*. Cambridge: Cambridge University Press, 297–299, 36, 72, 95, 302, 326ff.

Matsumoto, David., Frank, Mark G., and Hwang, Hyi Sung. (2013). *Nonverbal communication. Science and applications*. Los Angeles: Sage Inc., 34, 42, 78.

McKay, Matthew, Martha Davis, and Patrick Fanning. 2018. *Messages: The communication skills book*, 61F. Oakland: CA, New Harbinger Publications Inc.

Olekalns, Mara, Linda L. Putnam, Laurie R. Weingart, and Laurie Metcalf. 2008. Communication processes and conflict management. In *The psychology of conflict and conflict management in organisations*, ed. Carsten De Dreu and Michele J. Gelfand, 81–137. New York: Taylor & Francis Group.

Padilla Gálvez, Jesús. (2021). *El mentiroso. Genealogía de una paradoja sobre verdad y autorreferencia*. Valencia: Tirant humanidades, 340.

Powell, Mark. (2012). *International negotiations*. Cambridge, Cambridge University Press, 7ff.

Pruitt, Dean G., and Carnevale, Peter J. (1993). *Negotiation in social conflict*. California: Brooks/Cole Publishing, 31, 104–105, 123.

Uniform Commercial Code. (1952a). §§ 5–109. Fraud and forgery. *Legal Information Institute*. https://www.law.cornell.edu/ucc/5/5-109. 10 Sep 2023.

Uniform Commercial Code. (1952b). §§ 7–502. Rights acquired by due negotiation. *Legal Information Institute*. https://www.law.cornell.edu/ucc/7/7-502. 10 Sep 2023.

Uniform Commercial Code. (1952c). §§ 3–405. Employer's responsibility for fraudulent indorsement by employee. *Legal Information Institute*. https://www.law.cornell.edu/ucc/3/3-405. 10 Sep 2023.

Uniform Commercial Code. (1952d). §§ 2–271. Remedies for fraud. *Legal Information Institute*. https://www.law.cornell.edu/ucc/2/2-721. 10 Sep 2023.

Uniform Commercial Code. (1952e). §§ 3–404. Impostors, fictitious payees. *Legal Information Institute*. https://www.law.cornell.edu/ucc/3/3-404. 10 Sep 2023.

Wilson, Deirdre, and Dan Sperber. 1986. On defining relevance. In *Philosophical grounds of rationality*, ed. Richard E. Grandy and Richard Warner, 243–258. Oxford: Oxford University Press.

Zhou, Hui, and Tingqin Zhang. 2008. Body language in business negotiation. *International Journal of Business and Management* 3 (2): 90–96.

Chapter 6
Negotiation and Mediation

1 Introduction

In the first part of this chapter, we will present a definition of mediation and highlight the differences between negotiation and mediation. We will give an overview of the advantages of mediation in case of failed negotiation and conflict and examine mediation from the perspective of game theory. We discuss possible outcomes and refer to the interplay between mediation, forms of life and language games. The suitability of mediation regarding the resolution of deadlocked negotiations will be analysed.

The second part deals with procedural aspects of mediation. To understand the function and dynamics of mediation, we will examine typical features of the course of a conflict and how it leads to a stalemate in a negotiation. We present three models of mediation, transformative mediation, evaluative mediation and the *Harvard Negotiation Project* as a special form of alternative dispute resolution.

2 Definition of Mediation

Mediation is a negotiation-oriented process in which an impartial intermediary meets with negotiators to help the parties reach a mutual agreement. The reason why companies opt for mediation in case of an impasse in their business relations is that disputes are simply too costly for the companies involved. Prolonged conflicts cause a delay in business activity and loss of time and produce additional costs that are too expensive to sustain. This justifies the search for more efficient forms of dispute resolution.

Mediation is a form of alternative dispute resolution (ADR) because it is a simple and inexpensive way of reaching agreements without the need for court resolution. It is voluntary, confidential and collaborative, and there are binding and non-binding forms of ADR. Negotiation and mediation are generally non-binding, as they depend

M. Gaffal and J. Padilla Gálvez, *Dynamics of Rational Negotiation*,
https://doi.org/10.1007/978-3-031-49051-4_6

on the parties' willingness to reach an agreement. In contrast, arbitration can be binding or non-binding: in the first case, a third party proposes a solution that the parties must follow, and in the second case, the parties have the right to reject the decision. In its *Directive on Mediation in Civil and Commercial Matters*, the European Parliament has pointed to the advantages of mediation:

> 'Mediation' means a structured process, however named or referred to, whereby two or more parties to a dispute attempt by themselves, on a voluntary basis, to reach an agreement on the settlement of their dispute with the assistance of a mediator. This process may be initiated by the parties or suggested or ordered by a court or prescribed by the law of a Member State.[1]

Alternative dispute resolution has become an approved method of dispute resolution in many spheres of life. It enables parties to settle their disputes out of court and includes arbitration, mediation, neutral evaluation and conciliation.[2] As such, mediation is a first step in the dispute resolution process. From a cultural point of view, it can be considered a refined tool and an improvement to controversial judicial solutions that usually follow the "win-lose principle". Mediation is highly democratic in the sense that democracy can be seen as a continuous discourse towards consensual agreement.

In many countries mediation is widely practised in conflict resolutions and as a practical alternative to court resolutions. Mediation is also applied in public and private organizations to deal with interpersonal or institutional conflict. In conflicts between negotiating parties, mediation agreements are more appropriate than litigation. One reason for this is that mediation is based on cooperation and a holistic view of the institution or company in constant transition. The harmful consequences of conflict for companies and institutions must be kept in mind, as well as the high social and economic costs of litigation.[3] The consequences of conflict in the context of negotiations are particularly detrimental to third parties. Therefore, it seems advisable to introduce and promote mediation and to take appropriate measures to implement the use of mediation as a suitable means of resolving negotiation disputes.

If negotiations are difficult to initiate or have stalled, but the parties still want to avoid adversarial court proceedings, mediation can be a useful tool. C. Moore describes it as the assistance of a third party outside the dispute. He summarises the characteristics of mediation as follows:

---

[1] See: Article 3a. Definition. Directive (2008)/52/EC of the European Parliament and of the Council of 21 May 2008 on certain aspects of mediation in civil and commercial matters: https://eur-lex.europa.eu/eli/dir/2008/52/oj. (30/01/2023).

[2] As early as 1999, David Lipsky and Ronald Seeber conducted a survey in which they asked US companies whether they had already carried out alternative dispute resolution (ADR). The results of their survey showed that almost 90% of the companies had already used ADR. The reasons given for choosing this form of dispute resolution were cost savings and the possibility of resolving disputes on their own. Most companies rejected the court-imposed solution. Lipsky and Seeber (1999), 70–71.

[3] Lipsky and Seeber, citing cost savings as one of the reasons for resorting to ADR, indicated: "One of the more significant forces driving corporations towards ADR is the cost of litigation and the length of time needed to reach settlement. All else begun equal, ADR is widely considered cheaper and faster". Lipsky and Seeber (1999), 71.

> *Mediation* is an extension or elaboration of the negotiation process that involves the intervention of an acceptable third party who has limited (or no) authoritative decision making power. This person assists the principal parties to voluntarily reach a mutually acceptable settlement of the issues in dispute. As with negotiation, mediation leaves the decision-making power primarily in the hands of the people in conflict. Mediation is a voluntary process in that the participants must be willing to accept the assistance of the intervener if he or she is to help them manage or resolve their differences. Mediation is usually initiated when parties can no longer handle the conflict on their own and when the only means of resolution appears to involve impartial third-party assistance.[4]

In litigation, the aim of mediation is to reach a settlement even though the parties do not reconcile. It is customary for the parties to meet with the mediator if they have not been able to reach an agreement on their own. The mediator can deal with specific issues and supports the decision-making process between two parties whose interests do not coincide. The mediator assumes the role of a neutral third party without becoming a decision-maker or advocate for either party. In the European Directive, the role of the mediator is defined as follows:

> 'Mediator' means any third person who is asked to conduct a mediation in an effective, impartial and competent way, regardless of the denomination or profession of that third person in the Member State concerned and of the way in which the third person has been appointed or requested to conduct the mediation.[5]

Although mediation follows a set protocol, the parties' sense of fairness and mutual respect are equally important, as these are the basic conditions for agreement. However, it should be kept in mind that mediation is a non-legal dispute resolution process, which allows the parties to reach satisfactory solutions of their negotiations. While in most civil or administrative court cases the dispute between two or more parties revolves around financial issues, in institutional disputes immaterial issues are at stake. In the following section, we will approach mediation from the perspective of game theory and analyse the possible outcomes.

---

[4] Moore (2003), 8.

[5] See: Article 3b. Definition. Directive (2008)/52/EC of the European Parliament and of the Council of 21 May 2008 on certain aspects of mediation in civil and commercial matters: https://eur-lex.europa.eu/eli/dir/2008/52/oj. (08/01/2024).

Fig. 1 Game options

	Player 1 a	*Player 1b*
Player 2 A	2 / 2	0 / 0
Player 2 B	0 / 0	1 / 1

3 Mediation and Game Theory

Mediation can be seen as a shift of paradigm from the "win-lose principle"[6] to a "win–win solution".[7] While the former favours only one of the negotiating parties, the second provides benefits to all parties involved and avoids the exclusion of any of them. According to game theory, a "win–win strategy" is considered a cooperative approach, in which one party's gain does not build on the other's loss. In cooperative games, the parties are willing to cooperate, where group benefit is considered more important than individual gain or domination of one over the other.[8]

For the "win–win" strategy to develop, two preconditions are required: first, that there be contradiction of interests; and second, that the players be willing to cooperate with their interests. Figure 1 shows four possible options in an elementary game of collective benefit:

Options in a game in which all players win (a, b = strategies, 0/1/2 = amount of the winning margin). According to C. Rieck, there are four possible outcomes in a win–win position.[9] If both players choose the same strategy (a, A), then both will receive the maximum profit of 2 (option 1). This option represents cooperation between the two players, which produces a maximum gain for both. If player 1 chooses strategy a and player 2 chooses strategy B (a, B), or conversely, player 1 chooses b and player 2 chooses A (b, A), then cooperation fails and neither wins (options 2 and 3). If both players choose strategy (b, B), which represents cooperation under a commitment condition, then there is a restricted gain of 1 for the two players involved (option 4). This fourth option still produces winners, but with a reduced gain. Although human behaviour is much more complex than that described in the elementary win–win game, it nevertheless follows a similar logic. If the negotiators decide on the same strategy, the gain is maximised for both, but if their interests fundamentally clash,

[6] As we have seen in previous chapters, the win-lose model is closely linked to the zero-sum game, in which the dominant factor is that at least one person wins while another loses. Most two-person board games are also called zero-sum games. Win-lose games are based on a rational choice theory that maximizes deliberation and a broadly Hobbesian approach to bargaining.

[7] The bivalence system is applied by analogy to the game of chess: Poundstone (1992), 45–51. It describes a situation in which all players in the business win. It is also called a non-zero-sum game since everyone in the negotiation cooperates, is willing to make a compromise, or simply, everyone benefits because they cooperate.

[8] The application of the win–win principle shows how we relate our preferences to our actions. The strategy is guided by mutual-benefit optimality. In negotiation it is established that an objectively necessary agreement must be reached, determined by the idea of ex ante voluntary agreement.

[9] Rieck (1992), 58ff.

cooperation fails and gains are reduced. If both agree and reach the lowest common denominator, they still feel that they are winners, even if the payoff is reduced, but the possibility of a counter-reaction is low.

As we will see later, the Harvard Negotiation Project (HNP) defines win–win negotiations as integrative and lose-lose negotiations as distributive types of negotiation. Integrative negotiations are those in which the parties try to find a win–win solution by maximising mutual benefits.[10] If a party plans to maintain their business relation with a partner in the future, it seems prudent to use this form of negotiation because it allows for an amicable solution and helps improve the working relationship.[11] In contrast, in distributive negotiations, one party benefits at the expense of the other. This type of negotiation is not suitable for building long-term relationships.

As we have described above, it is more rewarding for negotiators to use a cooperative approach to conflict resolution. Through integrative negotiations, the difficulties associated with a court process are avoided, and with it the problem that internal company information, which is subject to confidentiality under normal business conditions, is disclosed by the court case. Even in the case of changing circumstance when parties need to readjust their interests and there are different individual preferences, a cooperative approach can turn into an advantage if the parties are willing to exchange mutual benefits. If the differences that the parties attach to different aspects of the negotiation can be identified, trade-offs can be achieved that bring benefits to both negotiators.[12] Thus, shared interests can turn a win-lose negotiation into a win–win deal.

From a social and political point of view, the zero-sum game produces exclusion and domination, while the win–win game enables collaboration and equality. Unfortunately, win-lose games are inevitable in our society and tend to have a destructive effect in interpersonal relationships.[13] From a psychological point of view, win-lose games arouse a spirit of opposition and provoke counterclaims, or else they result in submission and capitulation. Therefore, the great contribution of mediation to negotiation is that it replaces the traditional conception of win-lose with that of win–win. This change of perspective not only affects the results, but also the process itself, since it allows parties to change to reconsider and modify their attitudes.

---

[10] Pruitt and Rubin (1986), 139ff.

[11] Pruitt and Carnevale emphasize that if negotiations take place in a positive working relationship, the parties tend to make more concessions, show less hostile behaviour and cooperate more. Pruitt and Carnevale (1993), 136.

[12] Grice introduced the relevance maxim (Grice 1975, 64–75) which was systematized by D. Wilson and D. Sperber, indicating that relevance in each context is to be approached in terms of contextual implication (Wilson and Sperber 1986, 243–258).

[13] Win-lose games start from the dogma that individual actions are rational only if they are maximising. Maximization theory, in its various forms, is the default theory of rational choice in economics and political science. It is important to note that this proposition unravels important and surprising implications for our behaviour and, in particular, commitments to a Hobbesian approach to morality.

From a game-theoretic perspective, a cooperative approach helps maintain business relationships, whereas a distributive approach has a negative impact on negotiations. In the case of an adversarial negotiation, the parties tend to focus on maximising their gain, which increases tensions, as each party wants to win at the expense of the other. Mediation helps reduce these tensions by arbitrating between the parties. As it is based on the principle of the parties' voluntary participation, the mediator has no authority to impose a solution on the parties. The idea behind this principle is that resolutions that are reached by mutual compromise are more stable than solutions that have been imposed on one or both parties. Overall, the results show that mediation is a practical, less costly alternative to adversarial procedures. In fact, research has shown the compelling advantages that mediation has over disputes resolved in court. These advantages include greater satisfaction of negotiators with the process and its outcomes. Other advantages relate to the reduced costs of the entire procedure and a better adaptation rate in the post-negotiation situation.[14]

From a psychological point of view, self-negotiated agreements do not usually produce feelings of inequality, injustice or revenge. The mediation setting provides a safe and confidential space in which the parties can negotiate and reach an agreement. Unlike court settlements, mediation does not focus solely on the parties' legal rights, but allows them to express their views, which otherwise would have remained unspoken but which continue to exert their effect on the negotiating process and the business. This is a relevant criterion because it is often discordant views based on prejudices or emotions that play a decisive role in the development of conflicts. Prejudices or emotions linked to certain forms of life should never be disregarded as they play an important, sometimes decisive role in negotiation.

Most importantly, mediation helps improve communication and understanding between the negotiators, which is especially useful for the future relationship between the parties and for their success in the business. As mentioned above, mediation as a method of dispute resolution has many advantages over adversarial settlements. We have given a definition of mediation and distinguished it from other forms of interventions. We have then examined a summary of the advantages mediation offers. In the following sections, we will take a closer look at the role of negotiators. We start with the main actors, the negotiators, and then we will look at the role of the mediator. First, we want to look at who can really benefit from mediation. Does mediation make sense for all negotiations, or must the parties involved have certain predispositions or characteristics? We propose to answer these questions in the following section.

4 Suitability of Negotiators for Mediation

Mediation as a means of resolving a negotiation problem is not necessarily the best solution for all companies. It may only be successfully applied in businesses with certain characteristics. We may at the outset posit four interrelated criteria that we

[14] Gaffal (2010), 179 ff.

consider essential preconditions for the mediation process: first, each party's motivation for mediation; second, the parties' self-responsibility; third, their readiness to disclose their interests; and fourth, their willingness to reach a compromise. For the mediation to be successful, both negotiators must meet these criteria. Each negotiator should be motivated to participate in the mediation. The more sense of responsibility the negotiators bring to the table, the better the negotiation process will be. The parties should reveal their points of difference and put aside all prejudices. And most importantly, they should be willing to cooperate.

The basic attitude towards mediation depends on the parties' calculation of whether mediation can achieve more than a court decision.[15] A principle that applies to all types of conflicts is that if the parties want to avoid its escalation, the solution can only be found in continuing the negotiations or, if they stagnate, in mediation. The obstructive processes underlying negotiation disputes reveal similar characteristics to those observed in the emergence and development of political conflicts. It is therefore worth recalling briefly what the processes are that lead to the escalation of political conflicts and, in the worst case, to violence and war.[16]

We can define war as a state that emerges once conflicts become irresolvable and are followed by a cessation of negotiations, mutual threat and the declaration of war. This escalation can only end if the parties offer an armistice. To maintain the armistice, the parties must send intermediaries and mediators. At the negotiating table, they discuss solutions to end the conflict. If these negotiations turn out successful, a peace treaty is established, and measures are taken to maintain peace. While on the political level, the arbiters of these conflicts are the courts of law, in negotiating conflicts it is the mediator who intends to put an end to the dispute. The mediator acts as an intermediary who coordinates the negotiation process between the disputing parties. To manage the negotiations properly, the mediator proposes the most relevant points to be discussed and develops an agenda. As the manager of the negotiations, the mediator moderates discussion of the conflicting points to be resolved. The more coherent and organised the process, the easier it will be for the parties to arrive at mutually agreeable solutions.

Regarding the first criterion, it seems reasonable to assume that both parties should be motivated to enter mediation. However, a party's motive for choosing mediation as a method of dispute resolution often remains tacit. A motive may be the party's desire to avoid the negative consequences of a hostile legal battle, to save money, or to expedite the entire process of dissolving a contract. A party may also think altruistically and want to help the other party alleviate the consequences of contractual termination by reaching a mutual agreement. However, one should keep in mind the hidden and less beneficial reasons for mediation that could jeopardise the whole process. One such motive could be a party's intention to minimise the financial

[15] See the discussion on personal and social skills of negotiators in: Marrero-Ancízar et al. (2018), 8ff.

[16] Carl von Clausewitz's point of view, according to which all war conflicts are considered as the continuation of politics by other means, has become a typical cliché. Von Clausewitz (2021), 47f. Adapting this axiom to our research, we would say that every conflict is an effort by both sides to move their confrontation to a new level.

consequences of having initiated the dissolution of a contract. Other negotiators choose mediation in the hope of obtaining a better contract dissolution agreement by means of a power imbalance that favours them. Some enter the process seeking to reduce the consequences of their decision to terminate a contract and wishing to clarify responsibilities.

The second criterion, self-responsibility, is a fundamental precondition in which negotiators assume responsibility for the companies they work for.[17] At least three skills are associated with this criterion: the ability to overview the whole situation and its implications, knowledge of the company's priorities and its plans to achieve them, and the ability to clearly state its interests and goals and support these with rational arguments. Negotiators are faced with very basic questions, such as what their negotiating action is intended to achieve in the project being undertaken by the company. For any problem that may arise during the mediation process, it is the mediator's task to facilitate self-responsibility by providing essential information about the economic, practical and legal dimensions of a problem. Being a self-responsible negotiator can be a challenging task as it is linked to the form of life inculcated in a society. In a tense negotiation, lack of cooperation is often quickly seen by the other party as an obstacle to possible agreement.[18] However, if negotiations fail, the question of responsibility must be clarified: what proportion of the responsibility lies with the management and what proportion should be attributed to the negotiator?

The parties' readiness to disclose their interests as the third criterion in mediation requires that each party be asked to express its points of view openly. The aim is to avoid excluding issues that would subsequently impede the process. If hidden objections remain unspoken, they tend to lurk and resurface later. They would likely come to light later and obstruct the negotiating process. Rather, the danger lies in conflict avoidance and the tendency of mediators to try to sidestep controversy and disagreement.

Apart from the ability of a party to express its point of view, a fourth aspect necessary for mediation is the parties' willingness to reach an agreement. This marks a critical stage of the negotiation and is a challenge, as it requires a sense of agreement and tolerance of both negotiators and can easily be obstructed by opposing interests, objectives and methods.

From a cost–benefit point of view, it is useful for the mediator to identify in advance potential complicating factors that may impair the mediation process. Many negotiators are not interested in collecting financial data, but it is useful to have a model available to identify the economic pros and cons of possible outcomes. To assess whether or not the parties would benefit from mediation, the following five categories serve to describe the current state of negotiators in crisis: commitment

---

[17] This definition leaves aside those game strategies in which the negotiator prioritizes his personal gain, which is bound to generate conflicts.

[18] Marrero-Ancízar *et alii.* define accountability for agreements as follows: "responsible negotiators are sincere, fulfil their commitments reflected in the agreement, act ethically and honestly, are open to criticism and accept possible mistakes. On the contrary, a lack of integrity and responsibility would lead to the disapproval of the other party, creating attitudinal conflicts that make it impossible to reach a satisfactory agreement". Marrero-Ancízar et al. (2018), 57.

to negotiation, prior litigation, business relationship, communication skills, commitment to the company and the company's commitment to society. Although most negotiators would not fit exactly into these categories, the typology is a useful tool for decision-making.

Based on these categories we may identify altogether seven typologies of negotiators: collaborative negotiators, negotiators who block communication, negotiators that tend towards power struggles, negotiators with a tendency to give up easily, rival negotiators, confrontational negotiators and irascible negotiators. While with the first four types of negotiator it seems a priori possible to reach full or partial agreement, success is less likely for the last three typologies. We shall characterise the first four typologies in more detail and briefly describe the last three typologies of adversarial negotiators. These latter typologies have in common that their conflicts either have their origins in a particular form of life or have become too deeply rooted to be resolved in a reasonable time.

The following three typical characteristics of conflictual negotiators pose an obstacle to successful mediation. First, if there is too much tension, negotiators find it difficult to focus on an issue and relate to each other impartially. Second, an overall negative atmosphere during talks negatively influences the negotiators' readiness to agree on a solution with satisfactory long-term consequences. Third, the negotiators lack problem-solving strategies and do not know how to prevent an escalation of the conflict.

Negotiation is all about language, and if negotiators barely communicate, problems seem inevitable. Taciturn negotiators are typically introverted and find it difficult to express their interests and objectives. Negotiators of this kind are considered unprofessional, as they have not learned the appropriate language games used in their field of work.[19] This generates a deadlock in many negotiations, which makes mediation impossible as well. Some have interpreted such situations from a psychologist perspective, examining how a lack of linguistic training tends to create conflict.[20] A mediator could help in this case by introducing new ways of approaching the problem, teaching them how to express their interests and objectives. As such conflicts are mainly due to a lack of communication rather than personal dislike, a mediator can facilitate dialogue between the negotiators by adapting their language games in order to keep the communication flowing, which usually improves the situation.

The next typology has some similarities to the previous one in that negotiators struggling to get more out of a deal are usually wary and distrustful. Negotiators of this kind fear that the other might take advantage of any weaknesses they show and therefore compete on every detail of the deal. They use hostile communication and tend to adopt extreme positions so as not to reveal any weakness. In this case, the mediator will have to clarify the rules of mediation from the beginning by showing rigour rather than empathy and understanding. Although mediation seems much

---

[19] As we have seen in the previous chapters, the negotiator should know how to apply suitable language games, as not knowing the pertinent terminology will exacerbate the problems. Padilla Gálvez and Gaffal (2011), 9 ff. Padilla Gálvez and Gaffal (2013a), 12 ff. Gaffal (2016), 45 ff.

[20] See the study on the role of language skills in trade negotiations: Xiao et al. (2020), 186.

more difficult with this typology, it can lead to success if the mediator manages to help the negotiators out of the vicious circle of power struggle or the temptation to carry out camouflaged zero-sum strategies.[21]

While in the three previous typologies both negotiators want to undo the deal, there are other scenarios in which one negotiator wants to dissolve the contractual relationship and the other wants to prevent such dissolution by all means. Therefore, one negotiator will often tend to make unjustified or even impossible demands that may cause the conflict to escalate. Conversely, the opposing negotiator may be willing to make considerable concessions to convince the other negotiator to maintain the contract. Divergent views on the future of the relationship lead to an imbalance in desire to communicate. While the party who wishes to dissolve the relationship avoids communication, the party who wishes to maintain the relationship tends to take every issue as a starting point for discussion. In this case, the mediator must facilitate the transmission of messages between the negotiators.

Mediation is a much more difficult task in the case of adversarial negotiators because it ends the negotiation in a spiral of conflict from which they cannot extricate themselves without outside help. Negotiators often consult a mediator because they are either exhausted by the conflict or have been forced to undergo mandatory mediation. Although both negotiators know that the breakdown of contractual relations is inevitable, they are unable to reach an agreement when they both employ tactics for obtaining the better end of the deal. The negotiators barely communicate, and there is a danger that the companies will be drawn into the conflict. Although the prospects of reaching an agreement are relatively low, it is the mediator's task to find the common interest of both parties and to establish a regular schedule for the performance of contractual duties.

Similar conditions apply to entangled negotiations whose contractual dissolution is inevitable but neither negotiator assumes responsibility for the breakdown of discussions. Both are unaware of their problems but blame the other party for the situation. In this situation, there is little chance of success even for meditation and can often only be resolved by a court decision.

5 Characteristic Elements of Conflicts

Before presenting a theory on how agreements can be reached in negotiation conflicts, we need to know more about their structure and dynamics. Therefore, we will analyse their typical features as they manifest in negotiation.

A conflict can be defined as a state of opposition and discord between two or more parties caused by their real or perceived opposition of needs, values and interests. As such, a conflict denotes a situation in which the parties have incompatible objectives and seek to undermine each other's goals. In every conflict the parties involved choose, consciously or not, the steps that lead to its escalation or else to its solution.

[21] Von Neumann and Morgenstern 1944 (2004), 85 ss.

Assuming that there is a reasonable person participating in a conflict, whatever the person chooses will be the best possible course of action available to him or her at that time. Conflict can be seen as a crisis in human interaction. Thus, the crisis is a turning point at which the parties involved have to decide whether a course of action should be continued, modified or terminated.

Seen from this perspective, a conflict has characteristics similar to an iceberg, of which we can only see a part and the remaining, much larger part is invisible. The conflict that becomes manifest is only the top of the iceberg; however, the hidden parts play a decisive role in its development. According to Edward T. Hall, this mismatch between explicit and implicit conflict factors can be graphically represented by an iceberg.[22] Applying Hall's model to negotiations or mediation in companies one could say that what belongs to the visible part of a conflict are, for instance, the legal and contractual terms of the business relationship, the party's financial situation or the direct economic consequences of the negotiation. As such, a representation of the conflict shows its structure and short-term development. On the other hand, the submerged and implicit elements are unknown. Among these we can highlight the parties' interests, as we do not know whether there is a problem that affects the survival of the company.

Thomas and Kilman distinguish between the following five different styles that negotiators use to react to conflicts: accommodation, avoidance, collaboration, compromise or competition.[23] One party may want to accommodate the other party by capitulating and submitting its own needs. It may also avoid or postpone a conflict by ignoring it and thus gain time to concentrate on smaller, less important problems. Conflict avoidance may result in the cessation of communication or in one party keeping its distance. We also find that parties collaborate, working together to find a mutually beneficial solution. For collaboration to be effective, the parties must have a certain amount of mutual respect, trust and smooth communication. Compromise is achieved if the parties find a middle ground that partially satisfies both. Conflict can also lead to competition, in which one party asserts its interests at the expense of the other.[24]

The question is to know at what point the divergence between the parties becomes a conflict. In other words, what behaviours lead to conflict escalation and what behaviours lead to de-escalation? The author and conflict mediator F. Glasl has presented a general model that illustrates the steps on the way to conflict escalation. This model aims to facilitate conflict analysis and enables the parties or the mediator to develop strategies to counteract the conflict. The author gives the following definition of conflict:

> Sozialer Konflikt ist eine Interaktion zwischen Akteuren (Individuen, Gruppen, Organisationen usw.), wobei wenigstens ein Akteur Unvereinbarkeiten im Denken, Vorstellen, Wahrnehmen und/oder Fühlen und/oder Wollen mit dem anderen Akteur (anderen Akteuren)

---

22 Hall (1977), 57ff., 71ff.

23 Thomas and Kilman (2010).

24 For a detailed overview of the meta-strategies associated with each negotiation style see Meyer and Schoop (2023), 7, 11.

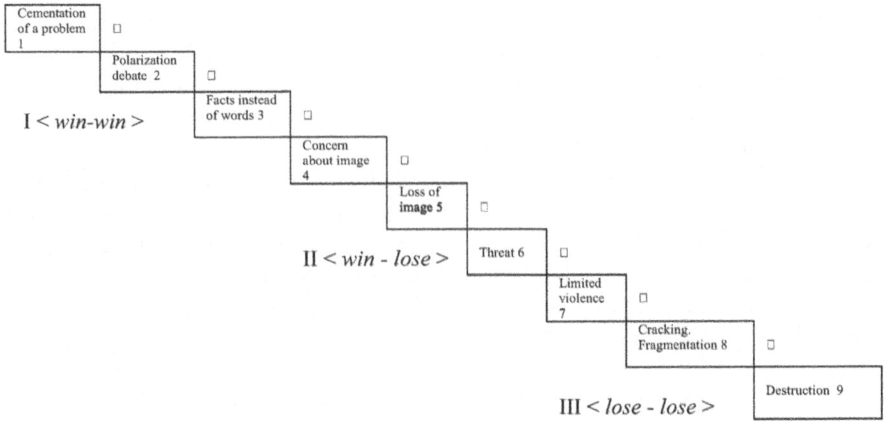

Fig. 2 Phases of conflict escalation

in der Art erlebt, dass im Realisieren eine Beeinträchtigung durch einen anderen Akteur (die anderen Akteure) erfolgt.[25]

Glasl's model demarcates three stages of human behaviour indicative of conflict, beginning with a win–win situation that develops into a win-lose situation and eventually deteriorates to a lose-lose position if no intervention is taken. Each of the three stages encompasses three sub-stages of typical behaviours that worsen a conflict. Figure 2 shows these nine stages of conflict escalation:

The first sign that indicates the deterioration of a winning position is the implementation of a problem (1st cementation). This is usually followed by polarisation and contentious debate (2nd polarisation and debate). Another critical point is reached if communication is affected and participants have stopped discussing an issue and started to act instead (3rd deeds instead of words). The third phase marks the first point of no return, because it means that communication has failed as a means of dealing with conflict. If words do not create an effect, the parties intend to put facts first. In contrast, in the "win–win" constellation, (steps 1–3) the parties still prefer and intend to reach a positive solution. In the next constellation, "win-lose" (steps 4–6), the conflict has taken a negative turn towards loss. At least from the end of step 3 onwards, the parties tend to follow the principle "As I cannot win, you must not win either".

At the beginning of the "win-lose" position, the participants are concerned about their image and the coalition at stake (4th concern about image and coalition). The parties try to force each other into fixed roles and positions and create stereotypes, image campaigns and rumours. At the same time, each seeks to gain adherents for its perspective. If a compromise cannot be reached, the parties try to damage the public image of the other party, which, as a reaction, provokes the fear of losing the image of both parties (5th loss of image). The parties initiate direct attacks with the

[25] Glasl (2004), 14 ff.

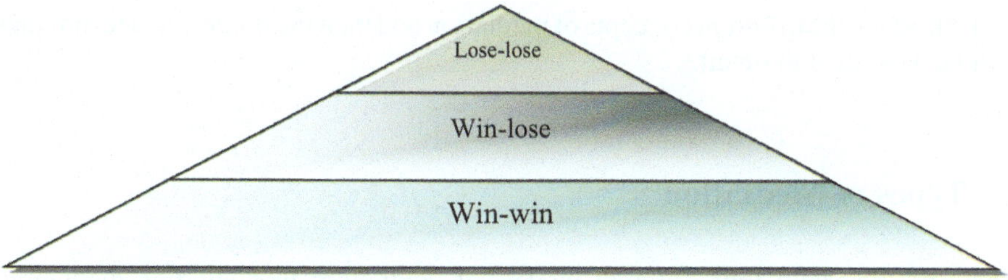

Fig. 3 Escalation of conflict between negotiators

aim of unmasking the other and thus provoking the observer's disappointment. The next stage is characterised by the threat, which is based on demanding the fulfilment of certain conditions and the obligation to reckon with the threats of the other party (6th threat strategies).

In the third stage of the lose-lose constellation, the conflict has gained momentum and the possibility of a positive solution is quite low. The conflict has become more stressful and has increased in speed. In phase 7, the parties use limited violence with the aim of forcing the opposing party to give in or reach a compromise. Strategically, the parties would present limited self-harm as a gain. In phase 8, one party seeks to destabilise the fundamentals, isolate certain groups from the centre of attention, and subsequently destroy the other (shattering). The parties attempt to destroy each other, either in the physical-material sense and/or on the mental-social plane. This phase becomes the ultimate point of no return, where both parties are at the edge of a precipice and will accept self-destruction if it means they can destroy the other party (destruction). Figure 3 shows the three stages and their constituent characteristics, depicting the escalation behaviour of the conflict between the negotiators:

For the mediator this is a useful scheme that helps to detect the phases of a conflict and develop de-escalation strategies. However, it must be kept in mind that in practice conflicts are more complex and their phases less linear than might be expected. Although human behaviour never occurs in a standardised way, the expert must be able to distinguish regularities and recurring patterns of behaviour. Thus, a mediator can help the parties understand their conflict and offer them conflict resolution strategies. As such, mediation can also be seen as a process in which the mediator guides the conflicting parties to undertake a series of constructive and well-founded steps. However, this is only possible if the negotiators accept the idea that a third person knows more about solving their conflict than they do.

At first glance, conflict tends to create feelings of instability and insecurity. However, it is not the conflict that is productive or destructive, but the way it is dealt with by the parties. The constructive aspect of conflicts and crises is an opportunity to balance inequalities and restructure personal, economic or political relationships. This allows conflict actors to adapt to the new conditions of the conflict framework and thus make progress. Conflict resolution is only possible if the parties see themselves as partners in conflict resolution and not as adversaries. In the following, we will introduce the different concepts of mediation and then the process and techniques will be described in detail.

6 Types of Mediation

There are basically four different models of mediation: first, guiding mediation; second, transforming mediation; third, evaluative mediation; and fourth, therapeutic mediation. We will only focus on the first three and leave aside therapeutic mediation, as it is not considered mediation in the strict sense because it is mainly applied in areas that affect the particular individuals involved in a conflict. Instead, we will introduce the Harvard Negotiation Project and describe the mediation concepts they apply. Note that mediators do not strictly adhere to a particular style but use a combination of these models to generate their own style.

Guided Mediation

In guided mediation, the mediator does not make recommendations or give advice on what the parties should do. Rather, his role is to ask questions, find the interests behind the parties' positions and help them find possible solutions. It is based on the belief that the people involved in the situation know best what they need for themselves. It is therefore a non-directive method in which the mediator supports the parties in making their own decisions. The mediator in guiding mediation structures a process to help the parties reach a mutually agreeable solution. The mediator asks questions, validates and normalises the views of the parties; seeks the interests underlying the decisions made by the parties and assists the parties in finding and analysing options for resolution. The guiding mediator does not make recommendations to the parties, give his or her own advice or option on the outcome of the case or predict what a court would do in the event it goes to litigation. The mediator is in charge of the process, while the parties are in charge of the outcome.

There are four characteristic elements to which mediation is geared: first, it is process-oriented; second, it is client-centred; third, it is communication-centred and fourth, it is based on the parties' interests. The essence of this form of mediation is that the mediator does not consider himself an expert in any of the fields associated with negotiation and refrains from making recommendations about the possible outcome. Rather, he gives the parties room to find their own solutions and only interferes in the mediation process itself.

Since the parties are fully responsible for the outcome and communication plays a crucial role, the mediator's main task is to help the parties communicate effectively. He or she will ask a party to clarify his or her point of view, inquire about the underlying interest or find out about the parties' preferred options. This method is based on the assumption that, once communication works, the parties will reach an agreement. The approach usually requires mediators to first establish effective communication between themselves and each of the parties. This sometimes means that direct communication between the parties is restricted until they are ready to engage directly with each other effectively.

The most important objective of guiding mediation is ascertaining the parties' interests and encouraging them to express them openly. It is assumed that joint understanding is a precondition for reaching an agreement and is facilitated once one party understands the interest of the other and appreciates the underlying need. The mediator will use an integrative approach and focus on the most reasonable options when dividing the limited resources available to them.

Transformative Mediation

Transformative mediation is a technique that helps clients to increase autonomy and self-responsibility and, at the same time, to become more responsive and respectful of other negotiators. In this case, mediation is defined as a process in which a third party works with the negotiating parties to help them change the quality of their conflict interaction from negative and destructive to positive and constructive as they explore and discuss the issues and possibilities for resolution.

Transformative mediation has two main objectives: empowerment and recognition. On the one hand, the mediator aims to empower the parties in conflict by helping them analyse the specific situation and respond appropriately to it. The overall objective is to support the parties' sense of responsibility and autonomy to make decisions independently. On the other hand, the mediator tries to improve the parties' understanding of and respect for the other negotiator by helping them see the opposing point of view. These two objectives constitute valuable interpersonal skills that are useful for communication in general. This enhances each party's appreciation of the other.

The mediator's primary objectives are to foster empowering changes by supporting, but never supplanting, each party's deliberation and decision-making, at every point in the session where choices arise regarding process or outcome; and to foster the acknowledgement of changes by encouraging and supporting, but never forcing, each party's freely chosen efforts to gain a new understanding of the other's perspective.

Proponents of transformative mediation assume that it is not only a method of conflict resolution, but can even bring about transformations in negotiators. This method is a tool to make negotiators more taxing and responsive to other negotiators. Mediators using this method will also place special emphasis on the party's responsibility for the outcome produced by the mediation. Accordingly, the mediator will basically tend to avoid any judgement on the views of the parties or the decisions made. Rather, the mediator takes an optimistic view and seeks to support the capabilities and resources of the parties.

The mediator's role is to help the parties make positive interactive changes—that is, empowering and acknowledging changes—by supporting the exercise of their driving and responsive capacities through deliberation, perspective-taking, communication and decision-making. In sessions, the mediator should give the parties space to express their concerns and emotions; the latter sometimes reveal implicit beliefs, so verbalising them can lead to a better understanding of underlying motives. The mediator should also encourage negotiators to talk about past events that created misunderstandings or confusion and try to bring clarity to the issue. Since past experiences are often the reason for a negotiator's reactions in the present, it is often helpful for a party to attempt to explain the reasons for his or her behaviour. However, the mediator must patiently allow each party explain and analyse its reactions rather than providing his or her own point of view. Clarification of the issues produces empowerment and recognition in the parties.

The mediator must listen carefully to detect and understand the message behind the client's words. Instead of giving an overview of what the client is saying, the mediator should concentrate on the statements made by a party. Mediators work in practice by translating these theoretical considerations into practical mediation skills. To do this they primarily use three techniques to achieve the goals of transformative mediation: attentive listening, reflection and summarising the most relevant points of the negotiation. Attentive listening implies that the mediator must be attentive to what the client says, as well as to what is implied by indirect non-verbal information such as gestures, intonation or non-verbal signs.[26] He or she must be able to detect the implicit information or "signifiers" behind the words used in the process.

The reflective technique is where the mediator reports what he hears the parties say and thus provides them with new issues to be discussed and analysed. In this way, the mediator encourages the parties to explain a point of view or to clarify vague or indistinct statements in more detail. In the reflective technique, the mediator talks to one client and allows the other client to listen from a distance to what the other client has said and then summarises in order for the mediator to talk to both clients. By summarizing the parts of a discussion or argument between the two parties, the mediator allows the parties to reconsider their positions. Other techniques include monitoring, questioning and silence. Generally, in transformative mediation, the emphasis is on considering the small steps taken by the clients. The mediator must share that success with his or her clients and adopt it as if were his or her own value before applying it to problem solving.

[26] Levinson has included among the non-verbal signals the gaze—that is, the eye contact between negotiators (Levinson 1997, 72, 95, 297 and 302)—the pauses (Levinson 1997, 299 f. and 326 ff.) and the prosody linked to intonation, lengthening and tension (Levinson 1997, 36).

Evaluative Mediation

Evaluative mediation has the following characteristics: first, the mediator's evaluation focuses on the state of the negotiations; second, it focuses on the behaviour of the parties; third, it channels the priorities of the parties; fourth, it specifically plans the resolution of the conflict and fifth, it develops alternatives outside the mediated negotiation.

Negotiators strive to achieve the maximum benefit for the party they represent and to achieve an optimal outcome for themselves. However, this strategy always involves the risk of losing the negotiation. This is not to create a negative image of negotiators, but simply to show how evaluative mediation can produce a more beneficial solution. In a mediated case, the negotiators can reach an agreement on how to finalise their negotiation through a contract that satisfies both parties in a reasonable manner, thus avoiding large losses. Sometimes one of the parties exhibits behaviour that leads to an impasse in the negotiations. In this case, the mediator can show the party in question the consequences of and possible alternatives to its conduct. This rethinking of a given position can take place in individual sessions to prevent the negotiators from losing face. In fact, mediation may be the only forum in which negotiators are willing to talk.

The mediator's goal is to ascertain the parties' priorities, as these are not always obvious at the outset. If the party's primary interest becomes apparent, negotiations will be more effective. A first step would be for the mediator to understand the priorities and then be able to evaluate them, always assessing what the possibilities for achieving the intended objective might look like. If a client has a concrete plan, the mediator can help evaluate this plan in the most effective and beneficial way.

The success of this technique depends largely on the experience the mediator has accumulated and the number of realistic estimates he or she can provide. Parties tend to achieve maximum benefit in these negotiations, and this benefit is always established in relation to other alternatives outside the negotiation. It is not surprising that a negotiator will refrain from resorting to mediation if he or she has a good chance of achieving a greater benefit with another negotiator. Therefore, it is important that the mediator has some experience and knowledge about trends in decisions and can give a realistic picture of the likelihood of success. It should be kept in mind that the evaluation should be geared towards encouraging acceptance of the proposed agreement, or considering the dangers of accepting the proposed terms that are always present in any mediation. The mediator's experience or expertise is very useful at that evaluative moment, whether that experience pertains to conduct in negotiations, finance or some other relevant area.

The Harvard Negotiation Project

In 1978, the *Harvard Negotiation Project* (HNP) was established as an interdisciplinary institute for research and teaching in negotiation and conflict management techniques at Harvard Law School. More specifically, law professors Roger Fisher and William Ury developed an outcome-oriented negotiation technique known as the "Harvard Negotiation Project".[27] They devised a series of theoretical concepts on successful negotiation and arbitration that could be used in many different fields, such as business, politics and many forms of out-of-court settlements. The method is based on a style of negotiation that has brought several novel features to the traditional culture of negotiation. It was designed to help people deal more effectively with different types of conflict.

The first thing they observed is that conflict parties too often focus on their winning strategies and neglect reaching a mutually acceptable and profitable agreement. This attitude is an obstacle to effective negotiation because it entrenches the parties' positions and tends to lead to stalemate. In such situations, the parties should rather be flexible and willing to continue negotiations.

If a failure of negotiations is foreseen, it is advisable to aim for the "best alternative to a negotiated agreement" (BATNA), which represents an agreement that can still be reached if the envisaged agreement proves impossible to achieve. This best alternative plays a crucial role in the negotiation project. The idea is that a party should not accept any solution worse than the best alternative to a negotiated agreement, although, in practice, this is often difficult to assess. It is assumed that each party has determined its individual best alternative considering the minimum acceptable outcome. Thus, the best alternative to a negotiated agreement is considered an open negotiating space bounded by a still acceptable outcome. The parties' assessment of their best alternative to a negotiated agreement will determine the minimum acceptable outcome of each of the negotiators in the negotiation, i.e. their withdrawal point (WP). The respective withdrawal points, in turn, will lead to the existence or not of a possible zone of agreement or, in other words, will delimit what is negotiable.

The negotiation technique used in the Harvard project is called "principled negotiations" because it is an approach that focuses on conflict management and resolution. Such negotiations are based on game theory and denote a win–win strategy.[28] This method is mostly used in the United States and has become a constructive tool for mediation practitioners. Before describing the different stages of the negotiation process, we will first present the five basic principles of out-of-court settlements proposed by the project. These principles are: first, flexibility and party autonomy; second, confidentiality and expert consultation; third, openness; fourth, rebuilding the relationship and fifth, cost avoidance.

The first principle of flexibility and party autonomy represents the freedom of the parties to choose the form of negotiation that suits them best and to set the outcomes accordingly. Although this autonomy is limited by national legal systems, the parties

---

[27] Fisher and Ury (2011).

[28] Von Neumann and Morgenstern, 1944 (2004).

still have considerable room for solutions. Whereas in legal settlements the individual interests of the parties are limited to legal positions, out-of-court settlements allow individual interests to take precedence over the legal formula. William Ury and Bruce Patton point out the advantage of being flexible in negotiations. They warn negotiating parties not to fall into the trap of negotiating over positions. They describe the unconstructive mechanism of positional bargaining, in which parties become locked into a position the more they try to clarify and defend it against objections. If a party concentrates primarily on its position, other possible solutions are set aside.[29]

Values, convictions and beliefs influence the positions taken by negotiating parties. Values are convictions that can be defined as strong persuasions or immovable beliefs without the need for proof or evidence. They are acquired during socialisation, through religion and customs and are transmitted mainly within the family. Once adopted, they are deeply anchored in the mental system and are often resistant to experience or objective evidence. These views cement our forms of life.[30] Conflicts are almost always linked to our values and convictions, especially if they relate to the form of life implicit in any society. When two negotiators enter a contract with the conviction that the deal should last as long as possible, its termination is an unexpected experience. However, it can be difficult for them to change their conviction because their business model was not designed to do so. This rigidity of beliefs and convictions makes mediation difficult because it leads to positional bargaining. Therefore, sometimes seemingly minor factors, such as the exposure of different forms of life, can play a crucial role in conflicts and hinder negotiations. After all, most of the problems we face are conflicts in which different forms of life are unconsciously accepted by the negotiating partners because they have been brought up in them. The difficulty of handling them lies essentially in the different perceptions of each of the people involved in the negotiation. When the discussion centres on a question of forms of life, the differences are practically irreconcilable and will grow, rather than diminish, with the contribution of arguments.[31] The more the discussion goes on, the more disparate the positions of the opposing parties will be.

In societies where the assumption of an objective reality is weak and beliefs and convictions are firmly rooted in the minds of negotiators, mediation is most likely to be a difficult task, because conflicts tend to end in positional bargaining. A further obstacle is that these convictions and beliefs are often part of a person's implicit knowledge, which creates a certain reluctance to discuss them openly. In this context, the mediator has the demanding task of making these convictions accessible for discussion. In the *Harvard Negotiation Project*, the authors have developed an alternative method to positional negotiation called principled negotiation. There are four cornerstones on which this method rests and they include the following instructions: first, detach people from the problem; second, focus on interests, not positions; third, create a range of possibilities before making a decision and fourth, insist that the outcome be based on some objective criterion.

---

[29] Fisher and Ury (1991), 5.

[30] Padilla Gálvez and Gaffal (2011, 2013, 9ff).

[31] Padilla Gálvez and Gaffal (2013b), 9ff.

The first requirement of detaching people from the problem points to the danger that conflicts tend to increase in complexity if they become personalised. It is therefore advisable for the participants to concentrate on the issue and avoid personal attacks. To do this, the parties have to accept that people see things differently and deal with conflicts differently. This is mainly due to the way we were brought up, socio-cultural influences and individual experiences, in short our forms of life. The parties must increase their mutual tolerance for each other's perspectives. Therefore, the parties should not only separate the person from the problem, but also focus on interests rather than positions. It is suggested that all those involved in the negotiations try to distance themselves from the conflict and try to analyse the underlying interests of the other party.[32]

From this new position the following question arises: what is the relevance of these criteria in mediating conflicts between two negotiators? Raising this question involves analysing the pragmatic elements of the language games involved in a negotiation and reviewing the circumstances in which they have been uttered.[33] It is certainly not easy for two negotiators who have reached an impasse in their negotiation to separate the person from the problem. Instead of asking the other what he or she wants, one must ask why he or she wants one thing or another. This is based on the assumption that everyone has a reason for a certain position and each interest points to a fundamental need. If we assume that the other party is not just acting arbitrarily or completely irrationally, we can always detect a need behind a certain interest. The mediator's task is to focus the parties' attention on these interests and to facilitate dialogue and communication on these issues. This can help the parties to understand each other better and to respect each other's interests.

The third element refers to inventing several possibilities before arriving at a decision. The idea behind this requirement is the degree of individual flexibility in finding solutions that are also acceptable to both parties. This requires a special effort, because we tend to be convinced of our own position and tend not to be creative in developing alternatives. In order to find alternatives, we have to overcome at least three obstacles, namely the avoidance of hasty judgements, the assumption that there are several solutions to the same problem, and the assumption that a solution depends exclusively on one party. The approach should be rather cooperative, in the sense that the parties should try to find common interests. The fourth tip refers to the claim that objective criteria should be introduced in the search for solutions to the problem. To avoid unreasonable tendencies, it is recommended that the parties agree on objective criteria before mediation begins. Keep in mind that the discussion of objective criteria reduces the number of compromises that each party must make and then unmake as they move towards agreement. In positional bargaining, negotiators spend much of their time defending their position and attacking that of the other party. People who

---

[32] Fisher and Ury (1991), 11.

[33] Pragmatic studies point to the importance of appropriate interpretation of relevant information. Wilson and Sperber have indicated that relevance in a given context is to be studied in terms of contextual implicature. Wilson and Sperber (1986), 243–258.

use objective criteria tend to spend their time more efficiently, talking about possible rules and solutions.[34]

In summary, the advantages of mediation include a number of both practical and emotional benefits. From a practical standpoint, negotiators maintain a greater degree of autonomy and have more control over the process and its outcome. Mediation follows a compact timetable, and the final agreement contains rules on how to deal with subsequent problems. Most importantly, mediation helps the parties accept their own needs and the justifiable rights and needs of the other. Finally, and most importantly, a hard-won agreement appears to be more durable and stable than a settlement inflicted on the parties. Subsequent litigation is therefore greatly reduced.

References

Fisher, Roger., Ury, William., and Patton, Bruce. (2011). *Getting to yes: Negotiating agreement without giving in*, [1981]. 3rd ed. New York: Penguin Books, 5, 11, 83.

Gaffal, Margit. (2010). Divorce mediation. In: *Psychosocial and legal perspectives of marital breakdown*. Berlin, Heidelberg: Springer, 179ff.

Gaffal, Margit. (2016). Actions embedded in forms of life. In: *Action, decision-making and forms of life*. Aporia, Vol. 9, 45ff. Berlin, Boston: De Gruyter.

Glasl, Friedrich. (2004). *Konfliktmanagement. Ein Handbuch für Führungskräfte, Beraterinnen und Berater*. Bern: Paul Haupt, 14ff.

Grice, Paul H. (1975). Logic and conversation. In: *The logic of grammar*, eds. D. Davidson and G. Harman. Encino, CA: Dickenson, 64–75.

Hall, Edward T. (1977). *Beyond culture*. New York: Anchor Books, 57ff., 71ff.

Levinson, Stephen C. (1997). *Pragmatics*. Cambridge: Cambridge University Press, 36, 72–95, 297–299, 302–326.

Lipsky, David B., and Ronald L. Seeber. 1999. Patterns of ADR use in corporate disputes. *Dispute Resolution Journal* 54 (66): 70–71.

Marrero-Ancízar, Yaimary., Ortiz-Torres, Maritza., Márquez-Sánchez, Fidel. (2018). Profile of trade negotiators: Elements for their characterization. In *GECONTEC: Revista Internacional de Gestión del Conocimiento y la Tecnología*, 6, 2, 50f., 57.

Mediation Directive. (2008). 2008/52/EC of the European Parliament and of the Council of 21 May 2008 on certain aspects of mediation in civil and commercial matters. *Official Journal of the European Union*. https://eur-lex.europa.eu/legal-content/EN/TXT/PDF/?uri=CELEX:32008L0052&qid=1694363213160. 10 Sep 2023.

Meyer, Marlene, and Mareike Schoop. 2023. Taxonomy of styles, strategies, and tactics in e-negotiations. In *Group decision and negotiation in the era of multimodal interactions*, vol. 478, ed. Y. Maemura, M. Horita, et al., 3–19. Cham: Springer.

Moore, Christopher W. 2003. *The mediation process: Practical strategies for resolving conflicts*, 8. San Francisco: Jossey-Bass.

Padilla Gálvez, Jesús., and Gaffal, Margit. (Eds.), (2011). *Forms of life and language games*. Frankfurt A. M., Paris, Lancaster, New Brunswick: Ontos Verlag, 9ff.

Padilla Gálvez, Jesús., and Gaffal, Margit (Eds.), (2013). *Forms of life and language games*. 2nd edition. Berlin: De Gruyter, 12ff.

Poundstone, William. (1992). *Prisoner's dilemma. John von Neumann, game theory, and the puzzle of the bomb*. New York: Doubleday, 45–51.

[34] Fisher and Ury (1991), 83.

Pruitt, Dean. G., and Rubin, Jeffrey Z. (1986). *Social conflict: Escalation, stalemate and settlement.* New York: Random House, 139ff.

Pruitt, Dean G., and Peter J. Carnevale. 1993. *Negotiation in social conflict*, 136. California: Brooks/Cole Publishing.

Rieck, Christian. 1992. *Spieltheorie. Eine Einführung*, 58–61. Eschborn: Christian Rieck Verlag.

Thomas, Kenneth W., and Ralph H. Kilmann. 2010. *Thomas-Kilman conflict mode instrument profile and interpretative report*. Sunnydale, CA: CPP Inc.

Von Clausewitz, Carl. (2021). *Vom Kriege*. Hamburg: Nikol Verlagsgesellschaft, 47f.

Von Neumann, John., and Morgenstern, Oskar. (1944). *Theory of games and economic behaviour.* Princeton: Princeton University Press, 2004, 85f.

Wilson, Deirdre, and Dan Sperber. 1986. On defining relevance. In *Philosophical grounds of rationality*, ed. Richard E. Grandy and Richard Warner, 243–258. Oxford: Oxford University Press.

Xiao, Pan., Luo, Xiaoyu., and Daly, Shawn, P. (2020). Language skills in business negotiation from the perspective of adaptation. *International Journal of Multidisciplinary and Current Educational Research*, 2, 4, 186.

Chapter 7
Conclusion

Negotiation is a common process and a suitable means that people use to reach agreement. It is used in many areas of life and at all levels: in family, among friends, at work, in the field of public administration, in politics, etc. Depending on the subject for negotiation, we try to harmonise our interests and goals with those of our negotiating partners. For daily interactions and informal agreements, each of us has a range of strategies available that we use in negotiations with others to reach a joint solution. Apart from these, professional negotiations are conducted by trained negotiators who are specialised in a particular field of work and who usually follow a pre-established protocol. This provides the framework within which an issue is discussed and where common positions are sought. Not only do negotiations occur in organisations, such as in corporations, companies, non-profit organisations or governments, but they also take place in legal proceedings. Examples of professional negotiations cover a wide spectrum, ranging from the fields of trade and commerce, diplomatic relations, political agreements to legal decisions and court orders.

The difficulty when studying negotiation is that it is a complex field of human interrelationships that requires the multidisciplinary study of several research areas. An investigation of the processes involved in negotiations includes mathematics, linguistics, psychology and intercultural studies. More specifically, in negotiation converge mathematical principles that relate to the field of game theory; linguistic analysis concerning the terminology of a subject area and the appropriate phraseology that negotiators use to present their interests and goals; psychological factors intervening in the context of persuasive communication and the intercultural knowledge necessary to avoid mistakes when negotiating or dealing with partners from other cultures.

The process of optimal decision-making by two or more negotiators with different objectives has been described in detail using the calculations within the framework of game theory. Game theory is a mathematical theory that attempts to resolve the various constellations that can arise during negotiations. Conflict is understood here

M. Gaffal and J. Padilla Gálvez, *Dynamics of Rational Negotiation*,
https://doi.org/10.1007/978-3-031-49051-4_7

in a general sense as a situation in which several participants hold different positions, and no agreement has yet been reached. It is an interactive situation in which several agents make decisions and whose outcome depends on the decision of all the agents involved. Each of the agents has its own preferences about the set of possible outcomes.

If we look at multicultural negotiations, in which people from different cultural backgrounds participate, language skills are an important tool for understanding. Undoubtedly, English is considered a *lingua franca* and is used by international business partners as a vehicle of communication and a common basis for the exchange of information. When negotiation takes place between intercultural partners, they are confronted with at least three challenges: first, they aim to seek agreement on their different interests; second, misunderstandings can arise on the linguistic level and third, they need to have intercultural skills to avoid any cultural mistakes. The appropriate and correct use of language plays a crucial role in international decision-making.

It can be particularly problematic if the negotiations come to a standstill because of the parties' divergent goals and they then blame each other for the stall. Thus, there are constructive models of negotiation and disadvantageous models that impede a solution. There is nothing more counterproductive than a negotiator's complaining about the other party's shortcomings. It is even more detrimental when such accusations are combined with cultural prejudice. As a result, the criticised negotiator automatically adopts a defensive position, developing the strategy of countercriticism based on the "tu quoque" fallacy. In return and as a part of this defence, the criticised partner accuses the other of having acted contrary to their principles. As such, the negotiators generate controversy without resolving the underlying problem. Negotiators are prone to use language games based on contempt and may even propagate false news or inaccurate information. The use of these techniques tends to destroy trust between negotiators and often leads to a loss in relations between companies and even between States. Rational negotiators usually stay away from these strategies and disapprove of this type of language, which is so harmful to negotiation. They seek to neutralise such negative processes by using appropriate language and by making solution-oriented suggestions.

After these preliminary remarks, we may define more precisely what we mean by negotiation. A negotiation is a dialogue between two or more persons or legal entities undertaken to achieve a consensus regarding certain issues. It is carried out via interaction between the negotiating parties, who intend to reach an agreement on the grounds of mutual interest. The agreement may be beneficial to all negotiating parties or only some of the parties involved. For establishing one's own requirements, it is important to recognise the needs and ambitions of the opposing party. The more clearly the parties' interests, goals and objectives are determined, the greater the possibility of reaching an agreement, pact or alliance. Negotiation may also be aimed at preventing conflicts, limiting certain risks or even containing possible disputes. Negotiations have a wide range of applications that span from minimising mutual losses to maximising gains. The objective of a negotiation can focuses on resolving points of difference, obtaining individual or mutual advantages or achieving results that satisfy multiple interests. Certain psychological aspects, such as negotiators' mutual trust, are essential factors for a negotiation to produce a positive outcome.

Although there is no consensus on whether we can establish a generic meaning that encompasses negotiation, there is nevertheless an important group of negotiation theorists who distinguish between distributive and integrative negotiation.[1] From our point of view, the linguistic skills and the mastery of certain language games determines what is understood by negotiation. In addition, multiple factors such as form of life, psychological constituents and negotiators' training and spatiotemporal circumstances must be considered. For example, one-off encounters are likely to generate distributive bargaining, whereas long-term relationships are usually conducted as integrative bargaining. In distributive bargaining, compromise, positional bargaining or hard bargaining, negotiators attempt to distribute a fixed number of benefits.

A distributive negotiation usually involves people who have never had a previous business relationship and are unlikely to do so again soon. The purchase of a car or a house is an example of distributive bargaining, as they operate under zero-sum conditions[2] where it is assumed that any gains made by one party will be at the expense of the other. In the negotiating process, each party usually takes an extreme position, which they know that it will not be accepted by the respective other. Whenever this situation arises, the parties offer to concede as little as possible before reaching an agreement. Distributive negotiators conceive of negotiation as a process of distributing a fixed amount of value that can only be reached by one at a time.

On the contrary, integrative negotiation is based on participants' interests, objectives or principles. It is based on a set of techniques that aim to improve the quality and likelihood of a negotiated agreement by considering how different parties value outcomes in different ways.[3] While distributive negotiation fixes the value to be shared between the negotiators, integrative negotiation creates value during the negotiation. To do so, integrative negotiators aim to offset losses against gains, known as "trade-offs" or "logrolling", or by reframing the conflict in such a way that the negotiators benefit from it in the form of a win–win negotiation.

Nevertheless, integrative negotiation has distributive elements, especially when the negotiating parties value certain elements equally or when details are left to be resolved at the end of the negotiation. While concessions are an elementary requirement in any negotiation, negotiators who give in quickly cut off the possibility of evaluating other integrative and mutually beneficial solutions. Consequently, the time factor is relevant when assessing the possibilities of an integrative negotiation. In fact, integrative negotiation involves a higher degree of trust and plays a more role in establishing lasting relationships. In addition, it allows for more efficient problem solving in the pursuit of mutual benefits. Integrative negotiation is based on the fact that an agreement provides an optimal gain in the negotiation rather than maximising the

[1] Walton et al. (1994).

[2] Von Neumann, Morgenstern, 1944 (2004).

[3] Von Neumann, Morgenstern, 1944 (2004). 98.

particular gain. Each negotiator allots to the other party a benefit sufficient to allow the agreement to be maintained.

Another bargaining model is linked to production. This type of negotiation focuses its strategy on the parties' underlying interests and approaches negotiation as a shared problem-solving exercise. The basis of productive negotiation is adherence to objective and principled criteria. What is called "integrated negotiation" is carried out by maximising the value in every single negotiation. This value maximisation is accomplished by linking and sequencing other negotiations and decisions related to a negotiator's operational activities. We shall explain this process by giving an example that resembles the pattern of the so-called *Peter Principle* as described in detail by Laurence J. Peter and Raymind Hull in their book The Peter Principle.[4] Let us assume that the conditions for a vacancy in a company are negotiated in such a way that a candidate with a higher profile than a rival candidate was ultimately not selected for the post. The commission wants to avoid the unexpected integration of a well-qualified candidate. The reason for this is that employees are not eager to improve the company's performance. Integrated bargaining can have a negative effect if, for example, a department consisting of mediocre colleagues prevents any improvement and thus rejects the better-qualified candidate. This tends to keep the overall performance of the company at a lower level.

The integrated negotiation approach in complex environments is carried out by mapping all potentially relevant negotiations, conflicts and operational decisions to integrate useful connections and minimise any potentially detrimental connections among them. In these complex environments, it may happen that a negotiator imposes his or her particular interest, as in the example of the vacancy described above. It is often observed that a negotiator or the department in question pretends to negotiate a different position or positions but has no intention of reaching an agreement and as a result ends up employing incompetent staff. This mode of bargaining is characterised by bad faith. This happens in the case of a company that opens a position but has no intention of making the necessary compromises to reach an agreement and leaves the positions vacant until a group of candidates has stabilised their positions. Negotiations in bad faith do not only occur in business but also take place in higher education and in politics. We are dealing with negotiation strategies in which there is no real intention to reach a compromise. The information processing model is always deficient. This model of insufficient or defective transmission of information is carried out through the abuse of certain psychological procedures. In decision-making, the transfer of information plays an important role. The opposing party is viewed as adversary simply because he could break the status quo and block someone from achieving his position or put into question any other particular interest. Any potential opposing negotiator is thus presumed to be implacably hostile, and counter-indicators are ignored. Consequently, the candidate's resume is undervalued by pointing to all contributions as signs of weakness. Acting in bad faith has enormous repercussions, as an institution can lose credibility where under-qualified and under-stimulated personnel dominate the working climate.

[4] Peter and Hull (1970), 80ff.

In the field of negotiation, the totality of advantages and disadvantages that are distributed in a negotiation is referred to as the negotiation pie. During the negotiation, the object of the negotiation may be fixed, increased or reduced. If the bargaining parties expand their bargain, then a win–win situation can arise, where all negotiators receive a piece of the pie. In practice, this maximisation approach is hindered by the small pie bias, since the negotiation parties have underestimated the size of the bargain. It may also happen that the increase of the pie is underestimated due to the incompatibility bias. In fact, the object to be bargained may be reduced during negotiations so that it leads to a reduction of the bargaining object, due, for example, to the fact that the costs of the negotiation influence the bargained-over object.

We have seen above that negotiators can introduce certain strategies that clearly allow them to maintain their own dominant position, sometimes to the detriment of the negotiation and its outcome. It seems appropriate to say a few words in this conclusion about the different ways of classifying the essential elements of negotiation. All negotiations are based on three basic elements: process, behaviour and substance. The process refers to the way in which the parties negotiate: the context of the negotiations, the parties to the negotiations, the tactics used by the parties and the sequence and stages in which all this takes place. The behaviour of the negotiators refers to the relationships between the parties, the communication between them, the implicit lifestyles and negotiation styles they adopt. The substance has to do with what the parties negotiate: the agenda, the issues—that is, the positions and interests—the options and the agreement or agreements reached at the end of the negotiation.

From another perspective, we have analysed negotiation from four different points of view: strategy, process, tools and tactics. Strategy is limited to the objectives that include the relationship between the parties and the results obtained in the negotiation. Processes and tools consider the steps to be followed and the roles to be adopted to prepare and negotiate with the other parties. Tactics encompass a very broad field of analysis of the statements and actions taken during the negotiation, as well as the responses and actions of the negotiators during the negotiation process. Some theorists add to this persuasion and influence, claiming that they have become integral to the success of negotiation today. Strategic approaches aimed at making certain concessions include consideration of the optimal time to make a concession, allocating certain concessions in quotas. It must be made clear that, in this process, the concession is not a re-evaluation of something already outlined. In addition, any negotiation must be aware of the cost of the concession.

All this has led to the conclusion that negotiation processes go through certain stages, which are systematised below. To understand the course of a negotiation, let us recapture the whole process: parties tend to seek information that confirms their own beliefs and tend to ignore information that contradicts their beliefs. To negotiate effectively, it is important to empathise with the other party's point of view and approach the issue from the other's perspective. A more direct approach to understanding the other party is to explicitly discuss each other's perceptions. Each person should openly share their perceptions without blaming or judging the negotiator in front of them. A negotiator may have prior perceptions and expectations about the other negotiator. To avoid prejudice, a party should convey their interest by allowing for an inclusive negotiation.

Careful listening is a crucial skill when negotiating. However, listening is different from hearing what the other party says. Active listening requires paying attention to what is said verbally and non-verbally during a negotiation. It may be necessary for the negotiators to clarify what they mean. By asking the respective other what he or she means, the negotiator will imply that the negotiation is not just following a routine but is being taken seriously. In any negotiation, too much information can be as harmful as too little. The principle of relevance helps us evaluate an excess of information, or a deficiency. Therefore, the negotiator must establish beforehand what he wishes to communicate to the other party before negotiating an important point. It is essential to determine the exact purpose of the information.

It may turn out during a negotiation that it is impossible to achieve the main objective. Therefore, negotiators should have an alternative available. The presentation of an alternative must be advantageous to both parties. This alternative process has been called best alternative to a negotiated agreement (BATNA). Its purpose is to improve the negotiation result of one of the parties. Alternatives must be real and feasible, and both parties can propose alternatives during a negotiation.

In the context of any negotiation, certain psychological aspects that directly concern the negotiators play an important role. For this reason, there has been talk of a mutual concern between negotiators in the resolution of conflicts. From this perspective, negotiators position themselves in the negotiation. On the one hand, there is concern for oneself and a concern for others. According to this model, negotiators balance their concern for personal needs and interests with the needs and interests of others. This approach can serve to clarify negotiators' preferences when negotiating a particular issue.

One of the most relevant factors when analysing negotiation is closely linked to the behaviour of negotiators during the process itself. To this end, different criteria have been introduced, ranging from considering negotiators in groups as "individualistic", "cooperative", "competitive" or "altruistic". They have also been labelled "soft", "hard" or "smart". There are also proposals that distinguish negotiators according to their "aggressive", "flexible" or "persuasive" attitude. In addition, we find divisions between "formalists", "cooperative", "diplomatic", "imposing" and "leading". Whatever epithets are used, the behaviour of the negotiator depends on many factors, which have been addressed in this book. However, a number of common traits can be indicated.

First, there is a group of negotiators who opt for moderate negotiation because they want to maintain relations with the competition. The offers they propose approve the demands of others, thus avoiding confrontation. The objective of negotiation is to reach an agreement, so they offer solutions by altering their points of view if necessary. Their opposite is the intransigent negotiator who applies contentious strategies in order to wield influence. Their negotiations are carried out through pressure. They insist on maintaining a dominant position, so opposing negotiators are considered adversaries. They personify the negotiation in themselves, so their goal is success. A third group aims to conduct an integrative negotiation and therefore avoids any commitment to a particular position. Their negotiation revolves around the issue being addressed. They consider the intentions, motives and needs of the negotiators as subsidiary. They disassociate people from the issue. They study interests. The result is a focus on drafting an agreement or contract between the negotiating parties. Decisions are made regardless of subjective issues such as power, dominance, pressure, subjective interest or arbitrary decisions.

As we have seen, tactics are an important part of the negotiation process. Most of the time they are subtle, difficult to identify and used for multiple purposes. Tactics are generally applied in distributive negotiations and if the objective is to neutralise the highest possible value from the negotiation. There are many negotiation tactics. In this study, we have briefly introduced some commonly used tactics. The best known is the auction, whose design generates competition. The principle is simple: when several negotiators wish to acquire the same object, they are willing to bid for it, which raises the price of the object. We also find negotiators who pursue a set of conditions to the point where the other negotiating party must either accept or give up. We also find decoy tactics used, e.g. to pretend that a minor issue is very significant, and later in the negotiation the issue may be altered by a concession of greater real stature. Skulduggery also plays an important role in negotiation, when negotiators propose extreme measures, often in the form of bluffs, to intimidate the other side into giving them what they want. There are also negotiations in which several levels of decision-making authority are involved to allow more concessions each time the agreement passes through a different level of authority. This strategy means that each time the offer is passed to a decision-maker, the decision-maker adds another concession before closing the deal. Another common tactic is presenting the other negotiator with a deadline, forcing them to make a decision. This procedure uses time to pressure the other party. Negotiating tactics linked to non-verbal procedures are also used. Thus, when a negotiator makes a proposal, his opponent may betray a strong negative reaction such as gasping or a visible expression of surprise or shock. This reaction suggests to the opposing party that the offer or proposal is absurd, which is meant to get the other party to lower its aspirations. Experiencing a non-verbal reaction in negotiation is sometimes more credible than hearing a negotiator make a verbal expression.

There are dual negotiations. Therefore, the tactic to follow is to entrust the negotiation to two negotiators whose functions are clearly divided. One of them pursues the conclusion of the contract and highlights the positive aspects of the negotiation, thus describing the mutual interests; the other should be in charge of highlighting the negative aspects, i.e. the opposing interests. The distribution of the two roles to two different negotiators allows the one who promotes the contract to generate trust with the other party and to close the contract. In this same procedure, we find the strategy of presenting a ridiculously high or ridiculously low initial offer that is not achievable. The theory is that the extreme offer causes the other party to re-evaluate its initial offer and move closer to the middle ground. This strategy has the advantage that the party making the extreme demand appears more flexible when making concessions towards a more reasonable outcome. However, this tactic may provoke direct rejection by the opposing negotiator, as he may think that he is dealing with novices who lack experience and, therefore, that the negotiation is a waste of time. We also find excessive strategies in that so much information is presented that it is difficult to determine which information is important and which is insignificant. Many negotiators use technical language or jargon in negotiation to mask a simple answer. Negotiators also tend to imitate or mirror a negotiator's behaviour by repeating what the negotiator says. In this strategy, a negotiator repeats what the other negotiator has just said. This is intended to demonstrate attention paid to the issue being addressed in the negotiation and that a point has been acknowledged.

We have repeatedly referred to two different types of communication in the negotiation setting: verbal and non-verbal. Verbal communication is an effective negotiation that requires participants to effectively convey and interpret the information provided through specific language games. Finally, this information will be collected in text form. By this we mean the process of drafting the text of an agreement that all parties are willing to accept and sign. The negotiating parties may start with a draft text, consider suggestions and work to find a middle ground between several different positions. Examples of text-based negotiation include contract negotiations that end with the drafting of a contract. Verbal communication is aimed at finally fixing the text of such a relationship. On the other hand, non-verbal communication focuses on the study of body language and gestures. By understanding how non-verbal communication works, a negotiator is better equipped to interpret the information that other participants are communicating non-verbally, while keeping secret those things that would inhibit the ability to negotiate.

A negotiator can gain some leverage by first expressing his position and thereby create a dynamism of anchoring. In this way the party establishes a stance that will become the point of reference for the whole of the negotiation process. Similarly, certain advantages can be gained through non-verbal signals such as assuming a dominant posture, developing personal space, introducing a prima facie strategy. Thus, if a negotiator stands back and starts to speak more quietly, he can thereby give a signal of his superiority by withholding information from the negotiating partner. Apparently, any negotiator who heads the bargaining table is recognised as having a powerful position. As such, first impressions through gestures, gestures, eye contact or greeting can help to establish an advantageous starting position in any negotiation.

We have briefly indicated above the strategy followed in a negotiation involving two or more parties from the same negotiating group. In this strategy, a certain similarity can be observed in the language games used by the negotiators of a team. These language games are incorporated into the negotiation, so it is common for a similarity to emerge in the use of certain linguistic stereotypes between the negotiating parties.

In this book, we have systematically addressed all the aspects that influence human relations when carrying out negotiations. In this context, we have emphasised the language games they use for specific purposes, the actions to which the negotiations are directed, the degree of effectiveness of the corresponding strategies, the possibilities and impediments that may arise due to the forms of life that the negotiators share or do not share with each other. Complex models have been presented that come into play when two ideal negotiators interact with the purpose of acquiring or selling a product or service for specific purposes. Our experience in different countries has shown the inherent difficulty that exists between different forms of life, since their modes of interaction are fixed by specific language games that can promote negotiation but often interfere and impede it. We hope that this book can elucidate and clarify some problems related to negotiation and motivate the reader to investigate further.

Bibliography

Arrow, Kenneth J. 1951. *Social choice and individual values*. New York: Wiley.

Arrow, Kenneth J. (1983). Economic equilibrium (1968). In *Collected papers of Kenneth J. Arrow*, vol. 2, *General Equilibrium*, 107–132. Oxford: Basil Blackwell.

Baarslag, Tim. 2016. *Exploring the strategy space of negotiating agents: A framework for bidding, learning and accepting in automated negotiation*. Springer Theses: Springer Publishing.

Bacharach, Samuel B., and Edward J. Lawler. 1981. *Bargaining: Power, tactics, and outcomes*. San Francisco: Jossey-Bass Inc.

Barwise, Jon, and John Perry. 1983. *Situation and attitude*. Cambridge Mass: The MIT Press.

Batinic, Bernhard, and Markus Appel. 2008. *Medienpsychologie*. Berlin, Heidelberg: Springer.

Bazerman, Max H., and Margaret A. Neale. 1992. *Negotiating rationally*. New York: Free Press.

Bellman, Richard, and David Blackwell. 1949. Some two-person games involving bluffing. *Proceedings of the National Academy of Sciences* 35 (10): 600–605.

Berry, Bruce., and Fulmer, Ingrid S. 2006. Methodological challenges in the study of negotiator affect. In *Methods of Negotiation Research*, 19–164. eds. Peter Carnevale, and C. K. W. De Dreu. Leiden, Boston: Martinus Nijhoff.

Beth, Evert W. 1962. *Formal methods: An introduction to symbolic logic and to the study of effective operations in arithmetic and logic*. Dordecht-Holland: D. Reidel Publishing Company.

Borel, Émile. 1924a. Sur les jeux où interviennent le hasard et l'habilité des joueurs. In *Eléments de la théorie des Probabilités*, ed. J. Hermann, 204–224. Paris: Librairie Scientifique.

Borel, Émile. 1924b. *Éléments de la théorie des probabilités*. Paris: Hermann.

Borel, Émile., and André Chéron. 1940. *Théorie mathématique du bridge à la portée de tous*. Paris: Gauthier-Villars.

Borel, Émile. 1921. La théorie des jeux et les équations à noyau symétrique gauche, *Comptes rendus de l'Académie des sciences*, CLXXIII, 1304–1308. In: Borel, 1924b, 199–203.

Borel, Émile. 1927. Sur les systèmes de formes linéaires à déterminant symétrique gauche et la théorie générale des jeux. *Comptes rendus de l'Académie des sciences*, CLXXXIV, 52–53.

Borel, Émile. 1938. *Traité du calcul des probabilités et des applications. Applications aux jeux de hasard, rédigé avec la collab.* de Jean Ville. Paris: Gauthier-Villars.

Braithwaite, R.B., and F.P. Ramsey, eds. 1950. *The foundations of mathematics and other logical essays*. New York: Humanities Press.

Brams, Steven J. 2003. *Negotiation games: Applying game theory to bargaining and arbitration*, [1990]. 2nd ed. London, New York: Routledge.

Breslin, J. William., and Rubin, Jeffrey Z. (eds.) 1995. *Negotiation theory and practice*. Cambridge, MA: Program on Negotiation at Harvard Law School.

Bühler, Karl. 1965. *Sprachtheorie. Die Darstellungsfunktion der Sprache*. Stuttgart: Gustav Fischer Verlag.

Bühler, Karl. 2011. *The theory of language. The representational function of language*. Amsterdam; Philadelphia: John Benjamins Publishing.

Carroll, John W. 1987. Indefinite terminating points and the iterated prisoner's dilemma. *Theory and Decision* 22: 247–256.

Champsaur, Paul. 1975. Cooperation vs. Competition. *Journal of Economic Theory* 11: 394–417.
Church, Alonzo. 1936. A note on the *Entscheidungs problem. Journal of Symbolic Logic* 1: 40–41.
Cohen, Herb. 1980. *You can negotiate anything*. Secaucus, NJ: L. Stuart.
Davidson, Donald. 1980. *Essays on actions and events*. Oxford: Clarendon Press.
Davis, Morton. 1970. *Game theory: A nontechnical introduction*. New York: Basic Books.
Dawson, Roger. 2011. [1995]. *Secrets of power negotiating: Inside secrets from a master negotiator, 15th anniversary ed.:* Pompton Plains, NJ: Career Press.
Dennett, Daniel C. 1980. *Content and consciousness*. London, New York: Routledge.
Dolan, Simon L., and Kristine M. Kawamura. 2015. *Cross cultural competence: A field guide for developing global leaders and managers*, 301. Bradford: Emerald Publishing Limited.
Doyle, Arthur Conan. 1993. *The final problem* (1893). In *The great stories of Sherlock Holmes*, 576–589. London: Magpie Books.
DuFrene, Debbie D., ed. 2014. *Communication beyond boundaries*. New York: Business Expert Press.
Egan, Gerard. 1994. *Working the shadow side: A guide to positive behind-the-scenes management*. San Francisco: Jossey-Bass.
Ekman, Paul., and Wallace V. Friesen. 1969. Nonverbal leakage and clues to deception, Psychiatry. *Journal for the Study of Interpersonal Processes* 32 (1), 88–106.
Fisher, Roger, and Daniel Shapiro. 2005. *Beyond reason: Using emotions as you negotiate*. New York: Viking Press.
Fisher, Roger, Alan Sharp, and John Richardson. 1998. *Getting it done: How to lead when you're not in charge*. New York: Harper Business.
Fisher, Roger., and Brown, Scott. 1989. [1988]: *Getting together: Building relationships as we negotiate*. New York: Penguin Books.
Fisher, Roger., Ury, William., and Patton, Bruce. 2011. *Getting to yes: Negotiating agreement without giving in*, [1981]. 3rd ed. New York: Penguin Books.
Fodor, Jerry A. 1975. *The language of thought*. Cambridge Mass.: Harvard University Press.
Gaffal, Margit. 2010. *Psychosocial and legal perspectives of marital breakdown*. Berlin, Heidelberg: Springer.
Gaffal, Margit. 2021. Preferences in negotiations. *Dókos* 27–28: 53–69.
Gaffal, Margit. 2022a. Negociación, teoría de los juegos, juegos de lenguaje y formas de vida. *Dókos* 29–30: 101–110.
Gaffal, Margit. 2022b. La búsqueda de reglas en los juegos de lenguaje. Notas en los cuadernos de bolsillo. *Dókos* 29–30: 213–222.
Gaffal, Margit. 2010a. Divorce mediation. In *Psychosocial and legal perspectives of marital breakdown*, 173–212. Berlin, Heidelberg: Springer.
Gaffal, Margit. 2015. Form of life as a philosophical concept. In *Emergent forms of life in anglophone literature. Conceptual frameworks and critical analysis*, 27–39. eds. M. Basseler, D. Hartley, and A. Nünning. Trier: Wissenschaftlicher Verlag.
Gaffal, Margit. 2016. Actions embedded in forms of life. In: *Action, decision-making and forms of life*. Aporia, vol. 9, 45–59. Berlin, Boston: De Gruyter.
Gauthier, David. 1975. Reason and maximization. *Canadian Journal of Philosophy IV* 3: 411–433.
Gauthier, David. 2013. Achieving pareto-optimality: invisible hands social contracts, and rational deliberation. *Rationality, Markets and Morals* 4: 191–204.
Gauthier, David. 2022. *Rational deliberation: Selected writings*. eds. S. Dimock, Cl. Finkelstein, and Ch. Morris. Oxford: Oxford University Press.
Gesteland, Richard R., and Gesteland, Mary. 2010: *India – Cross-cultural business behavior. For business people, expatriates and scholars*. Denmark: Copenhagen Business School Press.
Gesteland, Richard R. 2012. *Cross-cultural business behavior. A guide for global management*. Frederiksberg: Copenhagen Business School Press.
Glasl, Friedrich, 2004. *Konfliktmanagement. Ein Handbuch für Führungskräfte, Beraterinnen und Berater*. Bern: Paul Haupt.
Grice, Paul H. 1989. *Studies in the way of words*. Cambridge Mass.: Harvard University Press.

Grice, Paul H., 1969: Utterer's Meaning and Intentions, *The Philosophical Review*, 78, 147-177. in: Grice, 1989, 86–116.

Grice, Paul H., 1975: Logic and Conversation, in: D. Davidson, G. Harman (eds.), *The Logic of Grammar*, Encino, CA: Dickenson, 64–75.

Grice, Paul H. 1978. Further notes on logic and conversation. In *Syntax and semantics*, vol. 3, 183–197. eds. P. Cole, J. Morgan. New York: Academic Press. Publicado in: Grice, 1989, 22–40.

Grice, Paul H. 1981. Presupposition and conversational implicature. In *Radical Pragmatics*, 183–198. ed. P. Cole. New York: Academic Press. In: Grice, 1989, 269–282.

Hall, Edward T. 1977. *Beyond culture*. New York: Anchor Books.

Hammond, John S., Ralph L. Keeney, and Howard Raiffa. 1999. *Smart choices: A practical guide to making better decisions*. Boston: Harvard Business School Press.

Harsanyi, John C. 1966. A general theory of rational behavior in game situations. *Econometrica* 34: 613–634.

Harsanyi, John C., and Reinhard Selten. 1987. *A general theory of equilibrium selection in games*. Cambridge, MA: MIT Press.

Harsanyi, John C. 1956. Approaches to the bargaining problem before and after the theory of games: a critical discussion of Zeuthen's, Hicks', and Nash's theories. *Econometrica* 24 (2), 144–157.

Hart, Sergiu. 1973. Values of mixed games. *International Journal of Games Theory* 2: 69–86.

Hobbes, Thomas. 2007. *Leviathan*. ed. Richard Tuck. Cambridge: Cambridge University Press.

Hornickel, Jim. 2013. *Negotiating success: Tips and tools for building rapport and dissolving conflict while still getting what you want*. Hoboken, New Jersey: John Wiley & Sons.

Howard, John V. 1988. Cooperation in the prisoner's dilemma. *Theory and Decision* 24: 203–213.

Huizinga, Johan. 2014. *Homo ludens: A study of the play-element of culture*. Mansfield Centre, CT: Martino Publishing.

Islam, Shafiqul., and Lawrence Susskind. 2012. *Water diplomacy: A negotiated approach to managing complex water networks*. RFF Press Water Policy Series. New York: RFF Press.

Jung, Stefanie, and Peter Krebs. 2019. *The essentials of contract negotiation*. Cham: Springer.

Kalai, Ehud. 1977. Proportional solutions to bargaining situations: interpersonal utility comparisons. *Econometrica* 45 (7): 1623–1630.

Kamlah, Wilhelm., and Paul Lorenzen. 1973. *Logische Propädeutik. Vorschule des vernünftigen Redens*. Mannheim, Wien, Zürich: Bibliographisches Institut.

Katz, Lothar. 2017. *Negotiating international business. The negotiator's reference guide to 50 countries around the world*. North Carleston, SC: CreateSpace.

Knight, Frank H. 2002. *Risk, uncertainty, and profit*. Washington, DC: Beard Books.

Kraines, David, and Vivian Kraines. 1989. Pavlov and the prisoner's dilemma. *Theory and Decision* 26: 47–79.

Kramer, Roderick M., and David M. Messick, eds. 1995. *Negotiation as a social process*. Thousand Oaks, CA: Sage Publications.

Kreps, David, Paul Milgrom, John Roberts, and Robert Wilson. 1982. Rational cooperation in the finitely-repeated prisoner's dilemma. *Journal of Economic Theory* 27: 245–252.

Kuhn, Steven. 2019. *Prisoner's dilemma. The Stanford encyclopedia of philosophy* (Winter 2019 Edition), ed. Edward N. Zalta. https://plato.stanford.edu/cgi-bin/encyclopedia/archinfo.cgi?entry=prisoner-dilemma. [12/03/2020].

Langholtz, Harvey J., and Chris E. Stout, eds. 2004. *The psychology of diplomacy. Psychological dimensions to war and peace*. Westport, CT: Praeger Publishers.

Lax, David A., and James K. Sebenius. 1986. *The manager as negotiator: Bargaining for cooperation and competitive gain*. New York: Free Press.

Lax, David A., and James K. Sebenius. 2006. *3-D Negotiation: Powerful tools to change the game in your most important deals*. Boston: Harvard Business School Press.

Leibniz, Wilhelm Gottfried. 1765. *Nouveaux essais sur l'entendement humain, Neue Abhandlung über den menschlichen Verstand*. Darmstadt: WB, 1985.

Levinson, Stephen C. 1997. *Pragmatics*. Cambridge: Cambridge University Press.

Lewicki, Roy J. 1986. Challenges of teaching negotiation. *Negotiation Journal* 2 (1): 15–28.
Lewicki, Roy J., Bruce Barry, and David M. Saunders. 2016. *[1997]: Essentials of negotiation*, 6th ed. New York: McGraw-Hill Education.
Lewicki, Roy J., D. M. Saunders, and B. Barry. 2010. *Negotiation*. Singapore: McGraw-Hill/Irwin.
Lewis, David K. 1969. *Convention*. Cambridge, MA: Harvard University Press.
Lewis, David K. 1979. Prisoner's dilemma is a newcomb problem. *Philosophy and Public Affairs* 8: 235–240.
Lipsky, David B., and Ronald L. Seeber. 1999. Patterns of ADR use in corporate disputes. *Dispute Resolution Journal* 54 (66): 66–71.
Locke, John. 1975. *1690: An essay concerning human understanding*. Oxford: Clarendon Press.
Lorenzen, Paul., and Kuno Lorenz. 1978. *Dialogische Logik*. Darmstadt: Wissenschaftliche Buchgesellschaft.
Malhotra, Deepak. 2016. *Negotiating the impossible: How to break deadlocks and resolve ugly conflicts, without money or muscle*. Oakland, CA: Berrett-Koehler Publishers.
Malhotra, Deepak, and Max H. Bazerman. 2007. *Negotiation genius: How to overcome obstacles and achieve brilliant results at the bargaining table and beyond*. New York: Bantam Books.
Mansbridge, Jane J., ed. 1990. *Beyond self-interest*. Chicago: University of Chicago Press.
Marcus, Leonard J., Barry C. Dorn, and Eric J. McNulty. 2011. *[1995]: Renegotiating health care: Resolving conflict to build collaboration*, 2nd ed. San Francisco: Jossey-Bass.
Marrero-Ancízar, Yaimary., Maritza Ortiz-Torres, and Fidel Márquez-Sánchez. 2018. Profile of trade negotiators: Elements for their characterization. *GECONTEC: Revista Internacional de Gestión del Conocimiento y la Tecnología*, 6 (2), 48–60.
Matsumoto, David., Mark G. Frank, and Hyisung Hwang. 2013. *Nonverbal communication. Science and applications*. Los Angeles: Sage Inc.
Mayberry, John P., John F. Nash, and M. Shubik. 1953. A comparison of treatments of a duopoly situation. *Econometrica* 21: 141–154.
McKay, Matthew, Martha Davis, and Patrick Fanning. 2018. *Messages: The communication skills book*. Oakland: CA, New Harbinger Publications Inc.
Menkel-Meadow, Carrie, and Michael Wheeler, eds. 2004. *What's fair: Ethics for negotiators*. San Francisco: Jossey-Bass.
Meyer, Marlene. and Mareike Schoop. 2023. Taxonomy of styles, strategies, and tactics in e-negotiations. In *Group decision and negotiation in the era of multimodal interactions*, Vol. 478, 3–19. eds. Y. Maemura, M. Horita et al. Cham: Springer.
Milinski, Manfred. 1987. TIT FOR TAT in Sticklebacks in the evolution of cooperation. *Nature* 325: 433–435.
Mnookin, Robert H. 2010. *Bargaining with the devil: When to negotiate, when to fight*. New York: Simon & Schuster.
Mnookin, Robert H., Lawrence Susskind, and Pacey C. Foster, eds. 1999. *Negotiating on behalf of others: Advice to lawyers, business executives, sports agents, diplomats, politicians, and everybody else. Sage Series on Negotiation and Dispute Resolution*. Thousand Oaks, CA: Sage Publications.
Mnookin, Robert H., Scott R. Peppet, and Andrew S. Tulumello. 2000. *Beyond winning: Negotiating to create value in deals and disputes*. Cambridge, MA: Belknap Press of Harvard University Press.
Rémond de Montmort, Pierre. 1708. *Essay d'analyse sur les jeux de hazard*. Paris: Jacque Quillau, Imprimeur-Juré-Libraire de l'Université.
Moore, Christopher W. 2003. *The mediation process: Practical strategies for resolving conflicts*. San Francisco: Jossey-Bass.
Morgenstern, Oscar. 1935. Vollkommene Voraussicht und wirtschaftliches Gleichgewicht. *Zeitschrift Für Nationalökonomie* 6 (3): 337–357.
Movius, Hallam, and Lawrence Susskind. 2009. *Built to win: Creating a world-class negotiating organization*. Boston: Harvard Business Press.

Myerson, Roger B. 1991. *Game theory: Analysis of conflict*. Cambridge Mass.: Harvard University Press.

Nash, John F. 1950a. Equilibrium Points in N-Person Games. *Proceedings of the National Academy of Sciences of the United States of America* 36 (1): 48–49.

Nash, John F. 1996. *Essays on game theory*. Cheltenham, Brookfield: Edward Elgar.

Nash, John F., and Lloyd S. Shapley. 1953. A simple three-person poker game. *Annals of Mathematics Studies*, 24. In: *Contributions to the theory of games*, Vol. I, 105–116. eds. H. W. Kuhn, and A. W. Tucker. Princeton: Princeton University Press. Republished in: Nash, 1996, 10–21.

Nash, John F. 1950b. The bargaining problem. *Econometrica* 18, 155–162. Republished in: Nash, 1996, 1–8.

Nash, John F. 1951. Non-cooperative games. *The Annals of Mathematics* 54(2), 286–295. Republished in: Nash, 1996, 22–33.

Nash, John F. 1953. Two-person cooperative games. *Econometrica* 21, 128–140. Republished in: Nash, 1996, 34–46.

Nash, John F. 1954. Some experimental n-person games. Republished in: Nash, 1996, 61–87.

Nielsen, Richard P. 1996. *The politics of ethics: Methods for acting, learning, and sometimes fighting with others in addressing ethics problems in organizational life*. New York: Oxford University Press.

Nierenberg, Gerard I., and Henry H. Calero. 2009. *[1968]. The new art of negotiating: How to close any deal*. Updated. Garden City Park, NY: Square One Publishers.

Nierenberg, Juliet, and Irene Ross. 2003. *The secrets of successful negotiation: Effective strategies to improve your negotiating skills*. London: Duncan Baird Publishers.

Olekalns, Mara, Linda L. Putnam, Laurie R. Weingart, and Laurie Metcalf. 2008. Communication processes and conflict management. In *The psychology of conflict and conflict management in organisations*, ed. Carsten De Dreu and Michele J. Gelfand, 81–137. New York: Taylor & Francis Group.

Osborne, Martin. 2004. *An introduction to game theory*. Oxford: Oxford University Press.

Padilla Gálvez, Jesús., and Margit Gaffal. (eds.) 2003. *Lengua de especialidad: Economía europea y derecho europeo/Language for Specific Purposes: European Economy and European Law/ Fachsprache: Europäische Wirtschaft und europäisches Recht*. Cuenca: Ediciones UCLM.

Padilla Gálvez, Jesús., and Margit Gaffal. 2005. *Spanienknigge. Sozioökonomische Einführung in die Interkulturalität*. München, Wien: Oldenbourg Wissenschaftsverlag.

Padilla Gálvez, Jesús., and Margit Gaffal (eds.) 2011. *Forms of life and language games*. Frankfurt A. M., Paris, Lancaster, New Brunswick: Ontos Verlag.

Padilla Gálvez, Jesús., and Margit Gaffal (eds.) 2012. *Doubtful certainties. Language-games, forms of life, relativism*. Frankfurt A. M.: Ontos Verlag.

Padilla Gálvez, Jesús., and Margit Gaffal (eds.) 2013a. *Forms of life and language games*. 2nd edition. Berlin: De Gruyter.

Padilla Gálvez, Jesús., and Margit Gaffal. (eds.) 2013b. *Formas de vida y juegos de lenguaje*. Madrid, México D.F.: Plaza y Valdés.

Padilla Gálvez, Jesús. (ed.) 2016. *Action, decision-making and forms of life*. Berlin, Boston: Walter de Gruyter Verlag.

Padilla Gálvez, Jesús. 2021. *El mentiroso. Genealogía de una paradoja sobre verdad y autorreferencia*. Valencia: Tirant humanidades.

Pareto, Vilfredo. 1920. *Compendio di sociologia generale*. Florence: Barbèra.

Pareto, Vilfredo. 1906. *Manuale di Economia politica con una introduzione alla scienza sociale*. Molano: società Editrice Libraria. Trad.: *Manual of Political Economy. A Critical and Variorum Edition*. ed. by Vilfredo Pareto, A. Montesa, A. Zanni, L. Bruni, J. Chipman, M. McLure. Oxford: Oxford University Press, 2014.

Parson, Talcott. 1937. *The structure of social action*. New York, London: McGraw-Hill Book Company, Inc.

Peter, Laurence J., and Raymond Hull. 1970. *The Peter principle: Why things always go wrong*, 80ff. London: Pan Books Ltd.

Poundstone, William. 1992. *Prisoner's dilemma. John von Neumann, game theory, and the puzzle of the bomb*. New York: Doubleday.
Powell, Mark. 2012. *International negotiations*. Cambridge: Cambridge University Press.
Poza García, David J., and Jose M. Galán Ordax, et.al. 2010. Emergencia de normas espaciales de negociación en el juego de demanda de Nash. In *4th International Conference on Industrial Engineering and Industrial Management XIV Congreso de Ingeniería de Organización*, 26–34. Donostia-San Sebastián.
Pruitt, Dean G., and Peter J. Carnevale. 1993. *Negotiation in social conflict*. California: Brooks/ Cole Publishing.
Raiffa, Howard. 1982. *The art and science of negotiation*. Cambridge, MA: Belknap Press of Harvard University Press.
Raiffa, Howard, John Richardson, and David Metcalfe. 2002. *Negotiation analysis: The science and art of collaborative decision making*. Cambridge, MA: Belknap Press of Harvard University Press.
Ramsey, Frank P. 1990. *Philosophical papers*. Cambridge, MA: Cambridge University Press.
Ramsey, Frank P. 1991. *Notes on philosophy, probability and mathematics*. Napoli: Bibliopolis.
Ramsey, Frank P. 1978. *Foundations. Essays in philosophy, logic, mathematics and economics*. Ed. D. H. Mellor. London: Routledge & Kegan Paul.
Rapaport, Anatol. 1962. The use and misuse of game theory. *Scientific American* 207 (6): 108–119.
Rieck, Christian. 1992. *Spieltheorie. Eine Einführung*. Eschborn: Christian Rieck Verlag.
Rousseau, Jean-Jacques. 2011. *El contrato social*. Madrid: Editorial Gredos.
Salacuse, Jeswald W. 2013. *Negotiating life: Secrets for everyday diplomacy and deal making*. New York: Palgrave Macmillan.
Kupfer Schneider, Andrea., and Christopher Honeyman. (eds.) 2006. *The negotiator's fieldbook*. Washington, DC: American Bar Association, Section of Dispute Resolution.
Schranner, Matthias. 2008. *Costly mistakes*. Leck: CPI Clausen & Bosse Books GmbH.
Schulte, Oliver, and James Delgrande. 2004. Representing von Neumann-Morgenstern games in the situation calculus. *Annals of Mathematics and Artificial Intelligence* 42: 73–101.
Searle, John R. 1969. *Speech acts*. Cambridge Mass.: Cambridge University Press.
Searle, John R., 1985: *Intentionality. An essay in the philosophy of mind*. Cambridge: Cambridge University Press.
Selten, Reinhard C. 1965. Spieltheoretische Behandlung eines Oligopolmodells mit Nachfrageträgheit. *Zeitschrift Für Die Gesammte Staatswissenschaft* 121 (301–324): 667–689.
Selten, Reinhard C. 1975. Reexamination of the perfectness concept for equilibrium points in extensive games. *International Journal of Games Theory* 4: 25–55.
Selten, Reinhard C. 1980. A note on evolutionary stable strategies in asymmetric animal conflict. *Journal of Theoretical Biology* 84: 93–101.
Shaheed, Fatima, Sarit Kraus, and Michael Wooldridge. 2015. *Principles of automated negotiation*. Cambridge, UK; New York: Cambridge University Press.
Shapiro, Daniel. 2016. *Negotiating the nonnegotiable: How to resolve your most emotionally charged conflicts*. New York: Viking Press.
Shapiro, Ronald M., Mark A. Jankowski, and Jim Dale. 2001. *[1998]. The power of nice: How to negotiate so everyone wins—Especially you!* Revised. New York: John Wiley & Sons.
Shell, G. Richard. 2006. *[1999]. Bargaining for advantage: Negotiation strategies for reasonable people*, 2nd ed. New York: Penguin Books.
Shubik, Martin, ed. 1971. *Game theory and related approaches to social behavior. Selections*. New York: John Wiley & Sons.
Shubik, Martin. 1982. *Game theory in the social sciences, concepts and solutions*. Cambridge, MA: MIT Press.
Spangle, Michael, and Myra W. Isenhart. 2003. *Negotiation: Communication for diverse settings*. Thousand Oaks, CA: Sage Publications.
Subramanian, Guhan. 2011. *Dealmaking: New dealmaking strategies for a competitive marketplace*. New York: W. W. Norton & Company.

Susskind, Lawrence. 2014. *Good for you, great for me: Finding the trading zone and winning at win-win negotiation*. New York: Public Affairs.

Susskind, Lawrence, and Jeffrey L. Cruikshank. 1987. *Breaking the impasse: Consensual approaches to resolving public disputes*. New York: Basic Books.

Susskind, Lawrence, Sarah McKearnan, and Jennifer Thomas-Larmer, eds. 1999. *The consensus building handbook: A comprehensive guide to reaching agreement*. Thousand Oaks, CA: Sage Publications.

Sycara, Katia, Michele J. Gelfand, and Allison Abbe, eds. 2013. *Models for intercultural collaboration and negotiation. Advances in group decision and negotiation*, vol. 6. Dordrecht, New York: Springer Verlag.

Thomas, Kenneth W., and Ralph H. Kilmann. 2010. *Thomas-Kilman conflict mode instrument profile and interpretative report*. Sunnydale, CA: CPP Inc.

Turing, Alan. 1936–7. On computable numbers, with an application to the *Entscheidungsproblem, Proceedings of the London Mathematical Society Series* 2 (42): 230–265.

Ury, William. 2000. *[1999]. The third side: Why we fight and how we can stop*. Revised. New York: Penguin Books.

Ury, William. 2007a. *[1991]. Getting past no: Negotiating in difficult situations*. Revised. New York: Bantam Books.

Ury, William. 2007b. *The power of a positive no: How to say no and still get to yes*. New York: Bantam Books.

Ury, William. 2015. *Getting to yes with yourself, and other worthy opponents*. New York: Harper One.

van Eemeren, Frans H. 2010. *Strategic maneuvering in argumentative discourse: Extending the pragma-dialectical theory of argumentation. Argumentation in context. Vol. 2*. Amsterdam; Philadelphia: John Benjamins Publishing.

Von Clausewitz, Carl. 2021. [2008]: *Vom Kriege*. Hamburg: Nikol Verlagsgesellschaft.

Von Mises, Ludwig. 2011. *La acción humana. Tratado de economía*. Madrid: Unión Editorial.

Von Neumann, John. 1928. Zur Theorie der Gesellschaftsspiele. *Mathematische Annalen* 100 (1): 295–320.

Von Neumann, John, and Morgenstern, Oskar. 1944. *Theory of games and economic behaviour*. Princeton: Princeton University Press, 2004.

Walton, Richard E., Joel Cutcher-Gershenfeld, and Robert B. McKersie. 1994. *Strategic negotiations: A theory of change in labor-management relations*. Boston: Harvard Business School Press.

Watkins, Michael D. 2006. *Shaping the game: The new leader's guide to effective negotiating*. Boston: Harvard Business School Press.

Wilson, Robert. 1978. Information, efficiency, and the core of an economy. *Econometrica* 46: 807–816.

Wilson, Deirdre, and Dan Sperber. 1986. On defining relevance. In *Philosophical grounds of rationality*, ed. Richard E. Grandy and Richard Warner, 243–258. Oxford: Oxford University Press.

Wittgenstein, Ludwig. 2019. *Tratado lógico-filosófico. Logisch-philosophische Abhandlung*. Edición crítica de TS 204, introducción y traducción de Jesús Padilla Gálvez. Valencia: Tirant humanidades.

Wittgenstein, Ludwig. 2021. *Investigaciones Filosóficas*. 2ª edición. Traducción del original alemán, introducción y notas de Jesús Padilla Gálvez. Madrid: Editorial Trotta.

Xiao, Pan, Xiaoyu Luo, and Shawn P. Daly. 2020. Language skills in business negotiation from the perspective of adaptation. *International Journal of Multidisciplinary and Current Educational Research* 2 (4): 181–187.

Zartman, I. William., ed. 1978. *The negotiation process: Theories and applications*. Beverly Hills, CA: Sage Publications.

Zeckhauser, Richard, Ralph L. Keeney, and James K. Sebenius. 1996. *Wise choices: Decisions, games, and negotiations*. Harvard: Harvard Business School Press.

Zermelo, Ernst. (1913). Über eine Anwendung der Mengenlehre auf die Theorie des Schachspiels. In *Proceedings of the Fifth International Congress of Mathematicians*, vol. II, 501–504. ed. Ernest W. Hobson, and Augustus E. H. Love. Cambridge: Cambridge University Press.
Zhou, Hui, and Tingqin Zhang. 2008. Body language in business negotiation. *International Journal of Business and Management* 3 (2): 90–96.

SPRINGER NATURE

GPSR Compliance

The European Union's (EU) General Product Safety Regulation (GPSR) is a set of rules that requires consumer products to be safe and our obligations to ensure this.

If you have any concerns about our products, you can contact us on ProductSafety@springernature.com

In case Publisher is established outside the EU, the EU authorized representative is:

Springer Nature Customer Service Center GmbH
Europaplatz 3
69115 Heidelberg, Germany

The manufacturer's authorised representative in the EU is Springer Nature Customer Service Centre GmbH, Europaplatz 3, 69115 Heidelberg, Germany. If you have any concerns regarding our products, please contact ProductSafety@springernature.com

Printed and bound by CPI Group (UK) Ltd, Croydon, CR0 4YY
01/10/2026
02225916-0006